INVERSE AND ILL-POSED PROBLEMS SERIES

Introduction to the Theory of Inverse Problems

Also available in the Inverse and Ill-Posed Problems Series:

Monte Carlo Method for Solving Inverse Problems of Radiation Transfer
V.S. Antyufeev

Identification Problems of Wave Phenomena - Theory and Numerics
S.I. Kabanikhin and A. Lorenzi

Inverse Problems of Electromagnetic Geophysical Fields
P.S. Martyshko

Composite Type Equations and Inverse Problems
A.I. Kozhanov

Inverse Problems of Vibrational Spectroscopy
A.G. Yagola, I.V. Kochikov, G.M. Kuramshina and Yu.A. Pentin

Elements of the Theory of Inverse Problems
A.M. Denisov

Volterra Equations and Inverse Problems
A.L. Bughgeim

Small Parameter Method in Multidimensional Inverse Problems
A.S. Barashkov

Regularization, Uniqueness and Existence of Volterra Equations of the First Kind
A. Asanov

Methods for Solution of Nonlinear Operator Equations
V.P. Tanana

Inverse and Ill-Posed Sources Problems
Yu.E. Anikonov, B.A. Bubnov and G.N. Erokhin

Methods for Solving Operator Equations
V.P. Tanana

Nonclassical and Inverse Problems for Pseudoparabolic Equations
A. Asanov and E.R. Atamanov

Formulas in Inverse and Ill-Posed Problems
Yu.E. Anikonov

Inverse Logarithmic Potential Problem
V.G. Cherednichenko

Multidimensional Inverse and Ill-Posed Problems for Differential Equations
Yu.E. Anikonov

Ill-Posed Problems with A Priori Information
V.V. Vasin and A.L. Ageev

Integral Geometry of Tensor Fields
V.A. Sharafutdinov

Inverse Problems for Maxwell's Equations
V.G. Romanov and S.I. Kabanikhin

INVERSE AND ILL-POSED PROBLEMS SERIES

Introduction to the Theory of Inverse Problems

A.L. Bukhgeim

UTRECHT · BOSTON · KÖLN · TOKYO
2000

VSP BV
P.O. Box 346
3700 AH Zeist
The Netherlands

Tel: +31 30 692 5790
Fax: +31 30 693 2081
vsppub@compuserve.com
www.vsppub.com

© VSP BV 2000

First published in 2000

ISBN 90-6764-319-X

All rights reserved. No part of this publication may be reproduced, stored in a retrieval system, or transmitted in any form or by any means, electronic, mechanical, photocopying, recording or otherwise, without the prior permission of the copyright owner.

Printed in The Netherlands by Ridderprint bv, Ridderkerk.

Contents

Introduction	1
Chapter 1. Inverse problems for difference operators	7
1.1. Inverse problem of spectral analysis for Jacobi matrices	7
1.2. Inverse problem for a difference equation with constant coefficients	16
1.3. The problems of determining a difference operator in nonstationary statement	25
1.4. Remarks and references	39
Chapter 2. *A priori* estimates and the uniqueness of solutions of integro-differential equations with operator coefficients	41
2.1. Estimates of the Carleman type and their connection with the uniqueness of solutions of inverse problems	41
2.2. Estimates for the Schrödinger equation with operator coefficients	54
2.3. Remarks and references	64
Chapter 3. Inverse problems for differential equations	65
3.1. One-dimensional inverse problem for the wave equation in linearized statement	65
3.2. The method of transformation operators	75
3.3. Uniqueness in multidimensional inverse problems in nonstationary and spectral statements	84
3.4. Remarks and references	90

Chapter 4. Volterra operator equations and their applications — 93

4.1. Volterra operator equations in scales of Banach spaces 93
4.2. Nonhyperbolic Cauchy problem for the wave equation 110
4.3. The problem of integral geometry in a strip 115
4.4. The inverse problem of variational calculus 123
4.5. Remarks and references 132

Chapter 5. Foundations of the theory of conditionally well-posed problems — 133

5.1. Conditional well-posedness 133
5.2. l_h-well-posedness of difference schemes 146
5.3. Variational methods of solution of l_h-stable difference schemes . . 154
5.4. Remarks and references 160

Chapter 6. Theory of stability of difference schemes — 163

6.1. Statement of the problem and the necessary conditions of finite stability 163
6.2. Basic estimates 173
6.3. Sufficient stability conditions 179
6.4. Estimates of l-stability up to the boundary 192
6.5. Convergence theorems 196
6.6. Finite stability of two-layer schemes of the canonical form 206
6.7. Conditions of stability in terms of the transition operator 213
6.8. Remarks and references 218

Appendix A — 219

Bibliography — 225

Introduction

The scape of the monograph covers three subjects which are united by the research method — the *a priori* estimates method in scales of Hilbert or Banach spaces. The first subject (Chapters 1–3) includes inverse problems for differential and finite-difference equations. Its general statement is as follows. We have a certain one-parameter group (semigroup) of transformations $U(t)$ of the space H into itself and an element $f \in H$. If the "projection" P of the trajectory $u(t) = U(t)f$,

$$g = Pu, \tag{1}$$

is known, it is required to reconstruct the original curve $u(t)$ or the whole group (semigroup) $U(t)$.

Such formulation has vivid geometric interpretation; however, it is too general. The first important specification consists in consideration of the group of unitary operators in the Hilbert space H. In this case, by the Stown theorem, the curve $u(t)$ is a solution of the Schrödinger equation

$$\frac{1}{i}\frac{du}{dt} = Au, \quad u(0) = f, \tag{2}$$

where $A = A^*$ is the generating operator of the group $U(t)$. Thus, the problem arises of determining the operator coefficient A of the equation by a certain projection $g = Pu$ of its solution. Problems of the form (1), (2) are called inverse problems.

Example 1. Suppose $H = L_2(\Omega)$, $\Omega = \{x \in \mathbb{R}^n \mid |x| < 1\}$, $A = \Delta + a(x)$, where Δ is the Laplace operator, and the functions from the domain of A satisfy the Neumann condition on the boundary $\partial \Omega$.

$$\left.\frac{\partial u}{\partial \nu}\right|_{\partial \Omega} = 0.$$

It is required to reconstruct the potential $a(x)$ and, consequently, the operator A, by the function $g(x,t) = u(x,t)|_{\partial\Omega\times\mathbb{R}}$. In this example, the operation of "projection", P, is reduced to calculation of the trace of the solution u on the boundary of the cylinder $\partial\Omega \times \mathbb{R}$.

Analogous inverse problems can be formulated for the parabolic equation $u_t = Au$ and the hyperbolic equation $u_{tt} = Au$ ($A = A^* \leq 0$).

The solution of the Cauchy problem (2) may be written in spectral terms of the operator A. For simplicity, we assume that the spectrum of A is discrete:

$$u(t) = \sum_{k=1}^{\infty} e^{i\lambda_k t} \langle f, e_k \rangle e_k,$$

where $\{e_k\}$ is the orthonormal basis consisting of the eigenfunctions of the operator A, λ_k are the corresponding eigenvalues ($Ae_k = \lambda_k e_k$), $\langle \cdot, \cdot \rangle$ is a scalar product in H. Therefore, in view of (1) we obtain

$$g(t) = \sum_{k=1}^{\infty} e^{i\lambda_k t} \langle f, e_k \rangle P e_k.$$

Therefore, if the function $g(t)$ uniquely determines the operator A, this also holds for the spectral data $\{\lambda_k, \langle f, e_k \rangle P e_k\}$. Of course, the uniqueness theorems for the inverse problem (1), (2) or for its spectral analogue can be proved under the corresponding conditions for the initial data f and the operators A and P. As for the function f, we must consider two cases. In the first case, f is a generalized function with support concentrated on the boundary of the domain in which the operator A acts:

$$\operatorname{supp} f \subseteq \partial\Omega. \tag{3}$$

In the second case, f is a smooth function with support in Ω

$$\operatorname{supp} f \subseteq \Omega \tag{4}$$

and the operator A is sought on $\operatorname{supp} f$.

In the one-dimensional spectral variant of Example 1, for $f = \sigma(x)$, where $\sigma(x)$ is the Dirac function, the uniqueness theorem was established by Marchenko (1950). One-dimensional inverse problems of such type were considered by many authors (see, for example, Levitan (1962, 1984), Marchenko (1972, 1977), Lavrent'ev, et al. (1982) and the bibliography therein). In the multidimensional case, the problems of the first type (3) were first investigated by Berezanskii (1958, 1965) both in differential and finite-difference statements. Further, some linearized statements were considered in Lavrent'ev *et al.* (1980) and Romanov (1984). Multidimensional inverse problems in the statement (4) were first investigated by Bukhgeim and Yakhno (1976), where the theorems of uniqueness in the small were proved. Uniqueness "in the whole" in nonstationary statement was established by Bukhgeim and Klibanov (1981), and the spectral variant was considered in Bukhgeim (1985). Before we set forth these main results in Chapter 1, we study the difference analogues of inverse problems. Short Chapter 2 is devoted to the weight *a priori* estimates of the Carleman type for the Schrödinger operator. To prove them, we use the standard scheme (see Hörmander, 1969, Chapter 8; 1986, Chapter 14, and the bibliography therein). The new subject of our consideration is the Schrödinger equation with operator coefficients. This allows us to consider both one-dimensional and multidimensional inverse problems in a similar way.

In Chapter 3, we first consider the linearization method and the method of transformation operators. Then the uniqueness theorems for inverse problem (1), (2), (4) are established. For the operator A we take elliptic operators of the second order. The proof of these theorems consists in reducing them to the Cauchy problem for integro-differential equations and subsequent application of the estimates of Chapter 2.

The second subject — the integral equations in scales of Banach spaces and their applications — is treated in Chapter 4. This subject is connected with the existence of solutions of integro-differential equations. These equations are obtained as a result of solution of inverse problems. In contrast to Bukhgeim (1983b), the main attention is concentrated on the problems in which the reduction to the abstract theorem of solvability is not so evident. In particular, it is the problem of obtaining the asymptotics of continuity modules of the extension of a solution to the equation

$$u_{tt} - \Delta u - a(x)u = f$$

to the exterior of the cylinder $|x| > 1$, $x \in \mathbb{R}^2$. Here the transformation operators become unexpectedly useful. They are constructed in Chapter 3. It is interesting that there arises the duality concept by means of which we establish the uniqueness of the solution of the problem of reconstructing the function by the known integrals of it along a two-parameter set of parabolas with the weight which is analytic with respect to some variables. In the last section of Chapter 4, we find the necessary and sufficient conditions for the existence of solution to a certain inverse problem of variational calculus in the class of analytic functions. This problem consists in finding the Lagrangian

$$L(x, y, y') = v^{-1}(x, y) \times (1 + (y')^2)^{1/2}, \quad v > 0, \ v_y > 0$$

knowing its action

$$\tau(x_0, x_1) = \min \int_{x_0}^{x_1} L(x, y(x), y'(x))\, dx, \quad x_0, x_1 \in (-s, s).$$

The minimum is taken over all smooth curves which connect the points x_0 and x_1.

The main difficulty in investigating the problems of uniqueness in inverse problems is connected with their ill-posedness. Nonlinearity, as a rule, plays a subordinate role.

The third subject — the problems of developing numerical methods for linear ill-posed problems and principles of their discretization — is considered in Chapters 5, 6. In Chapter 5 we give the necessary notation and definitions, and in Chapter 6, taking the abstract Cauchy problem as an example, we build the theory of stability of finite-difference schemes for this problem.

Recall that the notion of well-posedness for the problems of mathematical physics introduced by Hadamard (1932) includes three conditions: the uniqueness, stability, and existence of solutions in the corresponding functional spaces. If the equation

$$Au = f, \quad u \in D(A) \tag{5}$$

in the normed space X is the mathematical description of the corresponding linear problem, then, from the point of view of functional analysis, all three conditions are reduced to the condition of existence of the *a priori* estimate

$$\|u\| \leq c\|Au\|, \quad \forall u \in D(A). \tag{6}$$

Since there are many applied problems which do not satisfy condition (6) in suitable functional spaces, Tikhonov (1943) proposed that the domain of definition should be narrowed down to a certain compact set $M \subseteq D(A)$ to recover the stability of ill-posed problems of the form (5). This approach can be justified by the fact that in practice we usually have more information about the solution than it is provided by (5). For example, if we seek the solution in the space of continuous functions $C[0,1]$, then from physical considerations it is often known that

$$\|u\|_{C^1[0,1]} \leq m.$$

We know that the set of such functions is compact in $C[0,1]$. In the general case, if
$$M = \{u \mid l(u) \leq m\},$$
where l is a certain seminorm, then the stability condition may be represented as the following *a priori* estimate of interpolation type:
$$\|u\| \leq \varepsilon l(u) + c(\varepsilon)\|Au\|, \qquad \forall \varepsilon > 0, \quad \forall u \in D(l).$$

Various methods of obtaining such estimates are proposed in Chapter 5. These estimates are also used for the numerical solution of equation (5).

Intensive development of the theory of ill-posed problems started in the mid-fifties of our century. It was connected with the advent of computers and the start of their extensive application (see John, 1955; Pucci, 1955; Ivanov, 1962; Lavrent'ev, 1962; Ivanov *et al.*, 1978 and the references therein). The concept of regularization introduced by Tikhonov (1963) played a significant role in the development of the theory of ill-posed problems.

The stability of difference schemes for the ill-posed Cauchy problem with constant coefficients was first investigated by Chudov (1962) who used the method of Fourier transform for this purpose. He also pointed out the problem of building the stability theory which in volves the equations with variable coefficients (Chudov, 1967). This problem was solved by the author (Bukhgeim, 1983a; 1986a) on the basis of the notion of stability of a difference scheme for the functions with finite support and the development of difference variants of *a priori* weight estimates of the Carleman type. The solution of this problem is given in Chapter 6. Since the technique of obtaining the sufficient conditions of stability in the case of variable coefficients is too complicated, the corresponding part may be omitted on first reading. Such estimates are interesting for the substantiation of difference methods of solution of multidimensional inverse problems considered in Chapter 3. In the case of the constant self-adjoint operator coefficient, the sufficient condition of stability of the two-layer equation with weights
$$\frac{u_{j+1} + u_j}{\tau} - A(\sigma u_{j+1} + (1-\sigma)u_j) = f_j, \quad A \geq 0$$
has simple form: $\sigma\tau\|A\| < 1$. The convergence theorems can be established for the case where the weight σ is coordinated in a certain way with the error of the initial data.

Practically, the three parts can be read independently. For the simplicity of presentation, we formulated some results in the form which is far from the most general one. Exercises are provided in the end of almost every section. Some of them are simple and some are interesting for research. The author does not give the complete bibliography on the issues discussed in this book since it includes several thousand titles. For other statements of inverse problems and methods of their solution we refer the reader to Marchenko (1950, 1972, 1977), Levitan (1962, 1984), Lavrent'ev (1962, 1965), Krein and Prozorskaya (1963), Leibenzon (1966), Lattès and Lions (1967), Nizhnik (1973), Romanov (1973, 1984), Faddeev (1974), Morozov (1974), Beznoshchenko and Prilepko (1977), Anikonov (1978), Ivanov et al. (1978), Tikhonov and Arsenin (1979), Lavrent'ev et al. (1980, 1982), Marchuk (1980), Chadan and Sabatier (1977), Preobrazhenskii and Pikalov (1982), Androshchuk (1986), and the bibliography therein.

The author thanks T. A. Voronina, N. P. Zenkova, and L. P. Sherstyugina for their help. The author also thanks V. B. Kardakov, who has read the book and corrected some inaccuracies.

Chapter 1.

Inverse problems for difference operators

1.1. INVERSE PROBLEM OF SPECTRAL ANALYSIS FOR JACOBI MATRICES

In this section we obtain the algorithm of reconstruction of a finite or infinite Jacobi matrix by certain functionals of its spectral data. We begin with the statement of the direct problem. In the complex Hilbert space $l_2(0, N-1)$ of sequences $u = (u_0, u_1, \ldots, u_{N-1})$ with scalar product

$$\langle u, v \rangle = \sum_{j=1}^{N-1} u_j \bar{v}_j, \qquad u, v \in \mathbb{C}^N$$

we consider the difference operator

$$(Au)_j = a_{j-1} u_{j-1} + a_j u_{j+1} + b_j u_j, \quad j = 0, \ldots, N-1. \tag{1.1}$$

We assume that

$$u_{-1} = 0, \quad u_N = 0, \quad a_j \neq 0, \quad b_j \in \mathbb{R}, \quad j = 0, \ldots, N-1. \tag{1.2}$$

The operator A may be rewritten as the three-diagonal matrix

$$Au = \begin{pmatrix} b_0 & a_0 & 0 & 0 & \cdots & & 0 \\ a_0 & b_1 & a_1 & 0 & \cdots & & 0 \\ 0 & a_1 & b_2 & a_2 & \cdots & & 0 \\ \cdot & \cdot & \cdot & \cdot & & & \cdot \\ 0 & \cdot & & & & \cdot & a_{N-2} \\ 0 & \cdot & & \cdot & 0 & a_{N-2} & b_{N-1} \end{pmatrix} \begin{pmatrix} u_0 \\ u_1 \\ \cdot \\ \cdot \\ \cdot \\ u_{N-1} \end{pmatrix},$$

which is called *a Jacobi matrix* if $a_k \neq 0$. Since the matrix A is self-adjoint, it has N real eigenvalues λ_k, $k = 1, \ldots, N$ (taking into account their multiplicity) and N eigenvectors e^k, which form the orthonormal basis in the space $l_2(0, N-1)$. Any Jacobi matrix has a good property which consists in the fact that all its eigenvalues λ_k are different. Therefore, without loss of generality, we may set

$$\lambda_1 < \lambda_2 < \ldots < \lambda_N.$$

To prove this, we assume the contrary. Then a certain eigenvalue is associated with two eigenvectors v and w: $Av = \lambda v$ and $Aw = \lambda w$. Note that

$$v_0 \neq 0, \qquad w_0 \neq 0. \tag{1.3}$$

Otherwise, assuming first $u = v$ and then $u = w$ in the recurrent formula

$$(Au)_j = a_{j-1}u_{j-1} + a_j u_{j+1} + b_j u_j = \lambda u_j, \qquad j = 0, \ldots, N-1, \quad u_{-1} = 0, \tag{1.4}$$

we obtain $v = 0$ and $w = 0$. Since an eigenvector is determined up to a constant factor, we may assume that $v_0 = w_0 = 1$. Then, setting $u = v - w$, we have $u_{-1} = u_0 = 0$, and as far as $Au = \lambda u$, from the recurrent formula (1.4) it follows that $u = 0$, i.e., $v = w$.

Now we recall the algorithm of finding the eigenvalues and eigenvectors for the Jacobi matrix A. Suppose $u_j = p_j(\lambda)$ is a solution of the difference Cauchy problem

$$Au = \lambda u, \qquad u_{-1} = 0, \quad u_0 = 1, \quad \lambda \in \mathbb{C}. \tag{1.5}$$

Then from (1.1), (1.2), and (1.5) we successively determine

$$(Au)_0 = a_0 u_1 + b_0 u_0 = \lambda u_0,$$

that is,
$$u_1 = \frac{(\lambda - b_0)}{a_0}, \qquad (Au)_1 = \lambda u_1,$$

which implies
$$u_2 = \frac{\lambda u_1 - b_1 u_1 - a_0 u_0}{a_1} = \frac{(\lambda - b_1)(\lambda - b_0)a_0^{-1} - a_0}{a_1}$$

$$\cdots\cdots\cdots$$

$$(Au)_{N-1} = \lambda u_{N-1}$$
$$u_N = \frac{\lambda u_{N-1} - b_{N-1} u_{N-1} - a_{N-2} u_{N-2}}{a_{N-1}}.$$

Note that $u_N = u_N(\lambda) = p_N(\lambda)$ is the polynomial of degree N. Since $u_N = 0$ by condition (1.2), equating $p_N(\lambda)$ to zero and finding the roots of this polynomial we determine the eigenvalues λ_k and the corresponding eigenvectors

$$p(\lambda_k) = (p_0(\lambda_k), p_1(\lambda_k), \ldots, p_{N-1}(\lambda_k)), \quad k = 1, \ldots, N$$

of the operator A. Setting

$$e^k = \frac{p(\lambda_k)}{\|p(\lambda_k)\|}, \quad k = 1, \ldots, N, \tag{1.6}$$

we get the complete orthonormal system of eigenvectors of the operator A.

So, given A we construct its spectrum $\{\lambda_k\}$ and the basis of orthonormal eigenfunctions $\{e^k\}$. In other words, we have solved the direct problem of spectral analysis.

Conversely, given the spectrum $\{\lambda_k\}$ and the complete set of eigenvectors e^k we may reconstruct the operator A by the known formula of spectral expansion of a self-adjoint operator:

$$A = \sum_{k=1}^{N} \lambda_k Q_k \tag{1.7}$$

where Q is the operator of projection onto the vector e^k

$$Q_k u = \langle u, e^k \rangle e^k.$$

However, in our case, the operator A has a special three-diagonal structure and therefore it is natural to suppose that it may be reconstructed using only a part of the spectral information $\{\lambda_k, e^k\}$. We begin the investigation of this problem of spectral analysis with the case where the unknown elements stand only in the principal diagonal. This means that a_j are known and b_j are unknown. The simple example of the two operators

$$A_1 = \begin{pmatrix} 0 & 1 \\ 1 & 1 \end{pmatrix}, \quad A_2 = \begin{pmatrix} 1 & 1 \\ 1 & 0 \end{pmatrix}$$

with the same eigenvalues shows that it is not sufficient to know the spectrum $\{\lambda_k\}$ in order to reconstruct the elements b_j. We need some information about eigenvectors. The formula of expansion in the Fourier series with respect to eigenvectors

$$f = \sum_{k=1}^{N} \langle f, e^k \rangle e^k, \quad f \in l_2(0, N-1) \tag{1.8}$$

and formula (1.7) give us additional information on the following set of functionals of $e^k = (e_0^k, e_1^k, \ldots, e_{N-1}^k)$:

$$\varkappa_k = \langle f, e^k \rangle e_0^k = (Q_k f)_0, \quad k = 1, \ldots, N, \tag{1.9}$$

where f is a given vector.

In this case the following theorem holds.

Theorem 1.1. *Suppose* $f \in l_2(0, N-1)$, $f_j \neq 0$ *for all* $j = 0, \ldots, N-1$. *Then the spectral data* $S_f = \{\lambda_k, \varkappa_k\}$, $k = 1, \ldots, N$, *uniquely determine* $\operatorname{diag} A = \{b_0, b_1, \ldots, b_{N-1}\}$.

Proof. In terms of the eigenvectors $p(\lambda_k)$ constructed above and normed by the condition $p_0(\lambda_k) = 1$, formulas (1.8) and (1.9), in view of relation (1.6), become as follows:

$$f = \sum_{k=1}^{N} \langle f, p(\lambda_k) \rangle \, \|p(\lambda_k)\|^{-2} p(\lambda_k) = \sum_{k=1}^{N} \varkappa_k p(\lambda_k) \tag{1.10}$$

$$\varkappa_k = \langle f, p(\lambda_k) \rangle \, \|p(\lambda_k)\|^{-2}.$$

Since $Ap(\lambda_k) = \lambda_k p(\lambda_k)$, we have $Af = \sum_{k=1}^{N} \varkappa_k \lambda_k p(\lambda_k)$ or, more specifically,

$$(Af)_j = a_{j-1} f_{j-1} + a_j f_{j+1} + b_j f_j = s_j, \tag{1.11}$$

Chapter 1. Inverse problems for difference operators

$$s_j = \sum_{k=1}^{N} \varkappa_k \lambda_k p_j(\lambda_k), \qquad (1.12)$$

$$p_{j+1}(\lambda_k) = \frac{\lambda_k p_j - b_j p_j - a_{j-1} p_{j-1}}{a_j}. \qquad (1.13)$$

Since $p_0(\lambda_k) = 1$, we know the number s_0. Therefore, setting $j = 0$ in (1.11) and recalling the assumption that $f_{-1} = 0$, we obtain

$$a_0 f_1 + b_0 f_0 = s_0$$

and find b_0:

$$b_0 = \frac{s_0 - a_0 f_1}{f_0}.$$

Knowing b_0, by formula (1.13) we have

$$p_1(\lambda_k) = \frac{\lambda_k - b_0}{a_0},$$

as $p_{-1}(\lambda_k) = 0$ and $p_0(\lambda) = 1$. By $p_1(\lambda_k)$, from formula (1.12) we obtain the number s_1. Then the process is repeated: setting $j = 1$ in (1.11), we obtain

$$b_1 = \frac{s_1 - a_1 f_2 - a_0 f_0}{f_1}.$$

Knowing b_1, from formulas (1.13) and (1.12) we find $p_2(\lambda_k)$, s_2, and so on. As a result, the numbers b_j are calculated by the recurrent formulas

$$b_j = \frac{s_j - a_{j-1} f_{j-1} - a_j f_{j+1}}{f_j},$$

$$p_{j+1}(\lambda_k) = \frac{\lambda_k p_j - b_j p_j - a_{j-1} p_{j-1}}{a_j},$$

$$s_j = \sum_{k=1}^{N} \varkappa_k \lambda_k p_j(\lambda_k), \qquad p_{-1} = 0, \quad p_0 = 1.$$

Thus, it is not only uniqueness that is established, but also the algorithm of reconstruction of diag A by the spectral data S_f is obtained. The theorem is proved. $\square$

Since the number of elements of S_f is equal to $2N$ and the number of parameters of the matrix A is equal to $2N - 1$, the question arises whether we can reconstruct all elements of A. The following example shows that we cannot do this.

Example 1.1. Suppose $N = 2$. Then the operators

$$A\pm = \frac{1}{6(2\pm\sqrt{2})}\begin{pmatrix} 11\pm 6\sqrt{2} & 3\pm\sqrt{2} \\ 3\pm\sqrt{2} & 1 \end{pmatrix}$$

have the same spectral data

$$S_f = \{\lambda_1, \lambda_2, \varkappa_1, \varkappa_2\}, qqf = (1,1)$$

$$\lambda_1 = 0, \quad \lambda_2 = 1, \quad \varkappa_1 = -\frac{1}{6}, \quad \varkappa_2 = \frac{7}{6}.$$

Evidently, this nonuniqueness is connected with nonlinearity of the problem.

The next theorem shows that given two spectral data S_f and S_g, where f and g are independent in a certain sense, we can determine all elements of the operator A.

Theorem 1.2. *Suppose $f, g \in l_2(0, N-1)$ and for all $j = 0, \ldots, N-2$*

$$\det\begin{pmatrix} f_{j+1} & f_j \\ g_{j+1} & g_j \end{pmatrix} \neq 0, \quad |f_{N-1}| + |g_{N-1}| \neq 0. \qquad (1.14)$$

Then the spectral data

$$S(f,g) = \{\lambda_k, \langle f, e^k\rangle e_0^k, \langle g, e^k\rangle e_0^k\}$$

uniquely determine the operator A.

Proof. Denote

$$\varkappa'_k = \langle g, p(\lambda_k)\rangle \|p(\lambda_k)\|^{-2}, \quad s'_j = \sum_{k=1}^{N} \varkappa'_k \lambda_k p_j(\lambda_k).$$

Then

$$(Ag)_j = a_{j-1}g_{j-1} + a_j g_{j+1} + b_j g_j = s'_j. \qquad (1.15)$$

Since the numbers s_0 and s'_0 are given, setting $j = 0$ in (1.11) and (1.15) and taking into account (1.14) we uniquely determine a_0 and b_0. Knowing them, we obtain $p_1(\lambda_k)$, s_1 and s'_1. Thereupon the process is repeated recurrently. At step $j = N-2$ we determine the numbers a_{N-2}, b_{N-2} and, finally, for $j = N-1$, taking into account the condition $|g_{N-1}| + |f_{N-1}| \neq 0$, we find the number b_{N-1}. The theorem is proved. □

Chapter 1. Inverse problems for difference operators

To generalize Theorems 1.1 and 1.2 to the case of $N = \infty$, we reformulate them in terms of the spectral function $\sigma(\lambda)$ of the operator A. This function is defined as the piecewise continuous function with jumps $\|p(\lambda_k)\|^{-2}$ at the points of the spectrum λ_k:

$$\sigma(\lambda) = \begin{cases} 0, & \lambda < \lambda_1, \\ \|p(\lambda_1)\|^{-2}, & \lambda_1 \leq \lambda < \lambda_2, \\ \cdots \\ \sigma(\lambda_k) + \|p(\lambda_{k+1})\|^{-2}, & \lambda_{k+1} \leq \lambda < \lambda_{k+2}, \quad k+2 \leq N, \\ \cdots \\ \sigma(\lambda_N), & \lambda \geq \lambda_N. \end{cases}$$

To this end, we define the Fourier transform F corresponding to the operator A

$$F : f \in l_2(0, N-1) \to L_2(\mathbb{R}^1, d\sigma(\lambda))$$

by the formula

$$Ff = \hat{f}(\lambda) = \langle f, p(\lambda) \rangle = \sum_{j=0}^{N-1} f_j p_j(\lambda), \quad \lambda \in \mathbb{R}.$$

Then from equality (1.10) we obtain

$$f = \sum_{k=1}^{N} \hat{f}(\lambda_k) \|p(\lambda_k)\|^{-2} p(\lambda_k) = \int_{-\infty}^{\infty} \hat{f}(\lambda) p(\lambda) \, d\sigma(\lambda).$$

Since $p(\lambda_k) \|p(\lambda_k)\|^{-1} = e^k$ is an orthonormal basis in $l_2(0, N-1)$, we have

$$\|f\|^2 = \sum_{k=1}^{N} |\hat{f}(\lambda_k)|^2 \|p(\lambda_k)\|^{-2} = \int_{-\infty}^{\infty} |\hat{f}(\lambda)|^2 \, d\sigma(\lambda)$$
$$= \|\hat{f}\|^2_{L_2(\mathbb{R}^1, d\sigma(\lambda))}, \tag{1.16}$$

where the last equality is the definition of the norm in the Hilbert space $L_2(\mathbb{R}^1, d\sigma(\lambda))$. Evidently, the transformation $F : f \to \hat{f}$ is an isomorphism of the space $l_2(0, N-1)$ onto $L_2(\mathbb{R}^1, d\sigma(\lambda))$. In this case, in view of (1.16) it conserves the norm. Thus, F is a unitary operator. Given a vector $f \in l_2(0, N-1)$ and a spectral function $\sigma(\lambda)$ we construct the measure

$$d\rho(\lambda) = \hat{f}(\lambda) \, d\sigma(\lambda) = \langle f, p(\lambda) \rangle \, d\sigma(\lambda).$$

In terms of $d\sigma(\lambda)$, formula (1.12) takes the form

$$s_j = \int_{-\infty}^{\infty} \lambda p_j(\lambda)\, d\sigma(\lambda).$$

Therefore, by Theorem 1.1, for $f_j \neq 0$, $j = 0,\ldots, N-1$ the measure $d\rho(\lambda)$ uniquely determines diag A.

Now, consider the case of $N = \infty$. Suppose that in the Hilbert space $l_2(0, \infty)$ of infinite sequences $u = (u_0, u_1, \ldots)$ with scalar product

$$\langle u, v \rangle = \sum_{j=0}^{\infty} u_j \bar{v}_j$$

an operator A is defined by formula (1.1). Suppose also that, as before, $a_j \neq 0$, $u_{-1} = 0$ and the numbers b_j from diag A are real. Assume for simplicity that $|a_j| \leq e$ and $|b_j| \leq e$. Then A is a bounded self-adjoint operator in $l_2(0, \infty)$. We consider the projection operator

$$\pi_N u = (u_0, u_1, \ldots, u_{N-1}, 0, 0, \ldots)$$

and the corresponding operator $A_N = \pi_N A \pi_N$ which acts in the space $l_2(0, N-1)$ and has the form (1.1), (1.2). As we have proved before,

$$\langle u, v \rangle_{l_2(0,N-1)} = \langle F_N u, F_N v \rangle_{L_2(\mathbb{R}, d\sigma_N(\lambda))}$$

where F_N is the unitary Fourier operator constructed from the operator A_N and $\sigma_N(\lambda)$ is its spectral function. We can show that the spectral function $\sigma_N(\lambda)$ converges to the function $\sigma(\lambda)$ as $N \to \infty$. This function $\sigma(\lambda)$ is called *the spectral function of the operator* A. Analogously, the operator F_N converges to the unitary operator $F: l_2(0, \infty) \to L_2(\mathbb{R}, d\sigma(\lambda))$ and the polynomials $p_j(\lambda)$, $j = 0, 1, 2, \ldots$ form a complete orthonormal system in $L_2(\mathbb{R}, d\sigma)$. In the general case, suppose A is a self-adjoint operator of the form (1.1), maybe unbounded, and E_λ is its expansion of the unity. This means that E_λ is continuous from the right, monotonically increasing family of orthogonal projectors in $l_2(0, \infty)$ such that

$$A = \int_{-\infty}^{\infty} \lambda\, dE_\lambda.$$

In this case we may show that

$$\sigma(\lambda) = \langle E_\lambda \delta_0, \delta_0 \rangle,$$

where $\delta_0 = (1, 0, 0, \ldots)$ and again $p_j(\lambda)$ form a complete orthonormal system in $L_2(\mathbb{R}, d\sigma)$. The following theorems can be established analogously to Theorems 1.1, 1.2.

Chapter 1. Inverse problems for difference operators

Theorem 1.3. *Suppose $f \in l_2(0, \infty)$ and $f_j \neq 0$ for all $j = 0, 1, 2, \ldots$. Then the measure*

$$d\rho(\lambda) = \langle f, p(\lambda) \rangle \, d\sigma(\lambda)$$

uniquely determines diag A.

Theorem 1.4. *Suppose $f, g \in l_2(0, \infty)$ and*

$$\det \begin{pmatrix} f_{j+1} & f_j \\ g_{j+1} & g_j \end{pmatrix} \neq 0, \qquad j = 0, 1, 2, \ldots .$$

Then the measures

$$d\rho_1(\lambda) = \langle f, p(\lambda) \rangle \, d\rho(\lambda), \qquad d\rho_2(\lambda) = \langle (g, p(\lambda) \rangle \, d\rho(\lambda)$$

uniquely determine the operator A.

Remark 1.1. Theorems 1.3, 1.4 are true for arbitrary self-adjoint three-diagonal operators A with $a_j \neq 0$ for all $j = 0, 1, 2, \ldots$. For example, it is known that under the condition

$$a_j > 0, \quad b_j \in \mathbb{R}, \quad \sum_{j=0}^{\infty} \frac{1}{a_j} = \infty$$

the operator A is self-adjoint (more precisely, it has a self-adjoint extension; see Berezanskii (1965)).

Exercise 1.1. Substantiate the passage to the limit from $\sigma_N(\lambda)$ and F_N to $\sigma(\lambda)$ and F. (Directions: use the Helly theorems (Berezanskii, 1965)).

Exercise 1.2. Prove that if $a_j > 0$, $j = 0, \ldots, N-2$ the spectral function $\sigma(\lambda)$ uniquely determines the entire operator A. Obtain the explicit formulas which express a_j and b_j in terms of $d\sigma(\lambda)$. (Directions: orthogonalize the sequence of functions $1, \lambda, \ldots, \lambda^{N-1}$ with respect to the measure $d\sigma(\lambda)$). Consider also the case of $n = \infty$. This exercise shows that in the finite-difference case the spectral measure $d\sigma(\lambda)$ carries more information about the operator A than $d\sigma(\lambda) = \langle f, p(\lambda) \rangle \, d\sigma(\lambda)$ where $f_j \neq 0$ for all $j = 0, \ldots, N-1$.

Exercise 1.3. For what measure $d\sigma(\lambda)$ does there exist an operator A of the form (1.2) with $a_j > 0$ and $\sigma(\lambda)$ as its spectral function? Consider the cases of finite and infinite N separately. (Directions: Exercises 1.2 and 1.3 in the case of finite N are considered in Atkinson (1964) and in the case of $N = \infty$ they are considered in Berezanskii (1965)).

Exercise 1.4. Show that the Frobenius matrix

$$\Phi = \begin{pmatrix} -a_1 & -a_2 & \cdot & -a_{N-1} & -a_N \\ 1 & 0 & \cdot & 0 & 0 \\ 0 & 1 & \cdot & 0 & 0 \\ \cdot & \cdot & \cdot & \cdot & \cdot \\ 0 & 0 & \cdot & 1 & 0 \end{pmatrix}$$

is reconstructed uniquely from its eigenvalues $\lambda_1, \lambda_2, \ldots, \lambda_{N_r}$. Here each eigenvalue λ_i is written p_i times, where p_i is its algebraic multiplicity.

Exercise 1.5. What can we say about a self-adjoint $n \times n$ matrix $A = (a_{ij})$, $a_{ij} = a_{ji}$, if its spectrum consists of one real eigenvalue $\lambda = 1$? The same question is for the matrix A without the condition of self-adjointness. (Directions: see Problem 116 in Selected Problems (1977)).

Exercise 1.6. Is it possible to reconstruct a cyclic matrix

$$A = \begin{pmatrix} a_1 & a_2 & a_3 & \cdot & a_n \\ a_n & a_1 & a_2 & \cdot & a_{n-1} \\ a_{n-1} & a_n & a_1 & \cdot & a_{n-2} \\ \cdot & \cdot & \cdot & \cdot & \cdot \\ a_2 & a_3 & a_4 & \cdot & a_1 \end{pmatrix}$$

from its spectrum?

1.2. INVERSE PROBLEM FOR A DIFFERENCE EQUATION WITH CONSTANT COEFFICIENTS

In this section we consider the problem of determination of a difference operator P_m with constant coefficients by the solution of the homogeneous equation $P_m u = 0$ given at finite number of points of the integer lattice.

Chapter 1. Inverse problems for difference operators

Solution of this problem associates the results of Section 1 with inverse problems in nonstationary statement, which are investigated in Section 3. As is shown in the next section, this connection, on the one hand, leads to the refinement of Theorems 1.1–1.4. On the other hand, it allows us to prove some uniqueness theorems which are not evident from the point of view of the nonstationary approach.

We shall now give the statement of the problem.

Suppose that $Z_+ = \{0, 1, \ldots\}$ and $u(x)$ is defined in Z_+ and takes values in $\mathbb{C}$, i.e., $u : Z_+ \to \mathbb{C}$, where $\mathbb{C}$ is the complex plane. The shift operator H is defined by the formula

$$Hu(x) = u(x+1), \qquad x \in Z_+.$$

The equation

$$P_m u \equiv \sum_{k=0}^{n_i} a_k H^{m-k} u(x) = f(x), \qquad (2.1)$$

where $f : Z_+ \to \mathbb{C}$ is a given function and a_k are given complex numbers such that

$$a_0 = 1, \qquad a_m \neq 0 \qquad (2.2)$$

is called *a linear difference equation of order m with constant coefficients a_k*. The polynomial

$$P_m(\lambda) = \lambda^m + a_1 \lambda^{m-1} + \ldots + a_m$$

is called *the symbol of the operator P_m* defined by formula (2.1), and the equation

$$P_m(\lambda) = 0$$

is called *the characteristic equation of the difference operator P_m*.

Further we shall consider only the homogeneous equation (2.1), i.e., the case of $f \equiv 0$. In order to describe the structure of its general solution, we recall some definitions. The Wronskian of the system of m functions $u_1(x), \ldots, u_m(x)$ is the following determinant

$$D[u_1(x), \ldots, u_m(x)] = \begin{vmatrix} u_1(x) & . & u_m(x) \\ Hu_1(x) & . & Hu_m(x) \\ . & . & . \\ H^{m-1}u_1(x) & . & H^{m-1}u_m(x) \end{vmatrix}.$$

The functions $u_j : Z_+ \to \mathbb{C}$, $j = 1,\ldots,m$, are called *linearly dependent* if there exist complex numbers $c_1,\ldots,c_m$ independent of x such that $\sum_{j=1}^{m} |c_j| \neq 0$ and for all $x \in Z_+$

$$\sum_{j=1}^{m} c_j u_j(x) = 0. \qquad (2.3)$$

The functions $u_1,\ldots,u_m$ which do not satisfy this condition are called *linearly independent*.

Theorem 2.1. *If the functions $u_j : Z_+ \to \mathbb{C}$, $j = 1,\ldots,m$, are linearly dependent, then*

$$D[u_1(x),\ldots,u_m(x)] = 0, \qquad \forall x \in Z_+. \qquad (2.4)$$

Conversely, if (2.4) holds and, moreover,

$$D[u_2(x),\ldots,u_m(x)] \neq 0, \qquad \forall x \in Z_+$$

then the functions u_j, $j = 1,\ldots m$, are linearly dependent and the number c_1 in (2.3) is not equal to zero.

The importance of this notion in the theory of equation $P_m u = 0$ becomes clear due to the following theorem.

Theorem 2.2. *Suppose $u_1(x),\ldots,u_m(x)$ are linearly independent solutions of equation $P_m u = 0$. Then any other solution of this equation may be represented as follows:*

$$u(x) = c_1 u_1(x) + \ldots + c_m u_m(x),$$

where $c_1,\ldots,c_m$ are some constants.

The proof of Theorems 2.1, 2.2 is left to the reader.

We can state two direct problems for the equation $P_m u = 0$.

Problem 2.1 [The Cauchy problem]. *Find a solution of the equation $P_m u = 0$ by the given numbers*

$$u(0) = u_0, \quad u(1) = u_1, \quad \ldots, \quad u(m-1) = u_{m-1}. \qquad (2.5)$$

Substituting the data (2.5) into the equation $P_m u = 0$, we find $u(m)$, and further we recurrently obtain $u(x)$ for any integer x. In order to deduce

Chapter 1. Inverse problems for difference operators

the general formula it is convenient to pass from the equation $P_m u = 0$ to the equivalent system of the first order. To this end, we set

$$\xi_0 = (u_0, u_1, \ldots, u_{m-1}), \quad \xi_n = H^n \xi_0 = (u_n, u_{n+1}, \ldots, u_{n+m-1}),$$

$$A = \begin{pmatrix} a_1 & a_2 & \cdot & a_{m-1} & a_m \\ 1 & 0 & \cdot & 0 & 0 \\ 0 & 1 & \cdot & 0 & 0 \\ \cdot & \cdot & & \cdot & \cdot \\ 0 & 0 & \cdot & 1 & 0 \end{pmatrix}.$$

Then Problem 2.1 is transformed into the problem

$$\xi_{n+1} = A\xi_n,$$

ξ_0 is known and, consequently,

$$\xi_n = A^n \xi_0, \quad u(n) = \langle A^n \xi_0, e_1 \rangle,$$

where $e_1 = (1, 0, \ldots, 0)$ and $\langle \cdot, \cdot \rangle$ is a scalar product in $\mathbb{C}^m$.

Problem 2.2. *Find the general solution of the equation $P_m u = 0$.*

We shall seek the solution in the form $u(x) = \lambda^x$. Substituting the function λ^x into the equation $P_m u = 0$ and dividing by λ^x, we pass to the characteristic equation

$$P_m(\lambda) = \lambda^m + a_1 \lambda^{m-1} + \ldots + a_m = 0.$$

Suppose that all the roots of the characteristic equation are simple. In this case, there exist exactly m solutions of the equation $P_m u = 0$ of the form

$$\lambda_1^x, \lambda_2^x, \ldots, \lambda_m^x, \qquad (2.6)$$

where λ_i are the roots of the characteristic equation $P_m(\lambda) = 0$. It is easy to see that these solutions are linearly independent. Indeed, the Wronskian of the system of functions (2.6) is

$$(\lambda_1 \lambda_2 \ldots \lambda_m)^x \begin{vmatrix} 1 & 1 & \cdot & 1 \\ \lambda_1 & \lambda_2 & \cdot & \lambda_m \\ \cdot & \cdot & & \cdot \\ \lambda_1^{m-1} & \lambda_2^{m-1} & \cdot & \lambda_m^{m-1} \end{vmatrix} = (\lambda_1 \lambda_2 \ldots \lambda_m)^x \prod_{m \geq k > j \geq 1} (\lambda_k - \lambda_j) \neq 0$$

(2.7)

since $\lambda_k \neq \lambda_j$ for $k \neq j$ and $a_m = (-1)^m \lambda_1 \lambda_2 \ldots \lambda_m = 0$ by condition (2.2).

Deducing equality (2.7), we use the formula for the Vandermonde determinant. From Theorem 2.2 it follows that the general solution has the form
$$u(x) = c_1 \lambda_1^x + \ldots + c_m \lambda_m^x,$$
where $c_1, \ldots, c_m$ are constants.

Now, we state the problems which, in a certain sense, are inverse to Problems 2.1, 2.2.

Problem 2.3. *Given the sequence of numbers*
$$u(j) = u_j, \qquad j = 0, \ldots, n-1, \qquad (2.8)$$
find the difference operator
$$P_m = \sum_{k=0}^{m} a_k H^{m-k}, \qquad a_0 = 1,$$
of minimal order m and the function $u : \mathbb{Z}_+ \to \mathbb{C}$ satisfying condition (2.8) such that $P_m u = 0$. In other words, we need to find the simplest law of formation of the sequence (2.8) under the assumption that this law is described by a linear homogeneous difference equation with constant coefficients.

Example 2.1. The Fibonacci sequence $0, 1, 1, 2, 3, 5, 8, 13, \ldots$. It is easy to see that this sequence is described by the equation
$$P_2 u \equiv u(x+2) - u(x+1) - u(x) = 0.$$
Here $m = 2$, $a_1 = a_2 = -1$.

Problem 2.4. *Find an integer $m \geq 1$ and nonzero numbers $c_j, \lambda_j \in \mathbb{C}$ from the system of equations*
$$\sum_{j=1}^{m} c_j \lambda_j^k = u_k, \qquad k = 0, \ldots, n-1, \qquad (2.9)$$
where the numbers u_k are given and all λ_j different.

Evidently, Problem 2.4 can be reduced to Problem 2.3. Indeed, formula (2.9) provides the general form of the solution of the difference equation

Chapter 1. Inverse problems for difference operators

$P_m u = 0$ with simple characteristic roots λ_j. Finding the operator P_m by the numbers u_j, i.e., solving Problem 2.3, we determine the number m and the coefficients a_j, $j = 1, \ldots, m$. Solving this characteristic equation, we find the numbers λ_j, $j = 1, \ldots, m$. Then we uniquely determine the numbers c_j from the system (2.9) since its determinant coincides with the Vandermonde determinant composed of different numbers λ_j.

An essential part of solution of Problem 2.3 is contained in the following theorem. We denote by $\mathcal{P}_m$ the set of all difference operators of order $k \leq m$ with constant coefficients of the form

$$P_k = \sum_{j=0}^{k} a_j H^{k-j}, \qquad a_0 = 1, \quad a_k \neq 0.$$

Theorem 2.3. *For the function $u : Z_+ \to \mathbb{C}$ to satisfy the difference equation $P_m u = 0$ of minimal order m, the following conditions are necessary:*

$$D[u(x), u(x+1), \ldots, u(x+m)] = 0, \qquad \forall x \in Z_+ \qquad (2.10)$$

and there exists a number $x \in Z_+$, $x \geq 1$ such that

$$D[u(x), u(x+1), \ldots, u(x+m-1)] \neq 0. \qquad (2.11)$$

Conversely, if condition (2.10) holds and for some integer $x \geq 1$ condition (2.11) holds, then there exists a unique operator P_m such that $P_m u = 0$ in the class $\mathcal{P}_m$.

Proof. First, we prove the sufficient part. For the determinants

$$D_m(x) \equiv D[u(x), \ldots, u(x+m)]$$

it is known (Pólya and Szegö, 1964) that

$$D_m(x)D_{m-2}(x) = D_{m-1}(x)D_m(x+2) - D_{m-1}^2(x+1). \qquad (2.12)$$

By the assumption of the theorem, we have $D_m(x) \equiv 0$, which implies

$$D_{m-1}^2(x+1) = D_{m-1}(x)D_{m-1}(x+2). \qquad (2.13)$$

We now show that $D_{m-1}(x) \neq 0$ for any integer $x \geq 0$. Indeed, let p be the minimal number for which $D_{m-1}(p) = 0$. If $p = 0$, then from (2.13) it follows that $D_{m-1}(x) \equiv 0$, $x \in Z_+$. If $p = 1$, then from (2.13) it follows again that $D_{m-1}(0) \neq 0$, $D_{m-1}(1) \neq 0$. Suppose that $p \geq 2$. Then

$$D_{m-1}^2(p-1) = D_{m-1}(p-2)D_{m-1}(p) = 0,$$

i.e., $D_{m-1}(p-1) = 0$, which contradicts the fact that p is minimal. So, $D_{m-1}(x) \neq 0$, $\forall x \in Z_+$ and $D_m(x) \equiv 0$. Therefore, by Theorem 2.1 the functions $u(x), u(x+1), \ldots, u(x+m)$ are linearly dependent, the coefficients of $u(x)$ and $u(x+m)$ being nonzero since

$$D[u(x), \ldots, u(x+m-1)] \neq 0, \qquad D[u(x+1), \ldots, u(x+m)] \neq 0.$$

This implies that there exists an operator $P_m \in \mathcal{P}_m$ such that

$$P_m u \equiv u(x+m) + a_1 u(x+m-1) + \ldots + a_m u(x) = 0. \qquad (2.14)$$

Now we prove the uniqueness of the operator P_m. Indeed, if there exists another operator

$$P'_m u \equiv u(x+m) + a'_1 u(x+m-1) + \ldots + a'_m u(x) = 0 \qquad (2.15)$$

of order m, then, subtracting equation (2.15) from equation (2.14), we find that the functions $u(x), \ldots, u(x+m-1)$ are linearly dependent. This contradicts the condition $D[u(x), \ldots, u(x+m-1)] \neq 0$. Therefore, $P'_m = P_m$ and sufficiency is established.

We prove the necessary part. Suppose $P_m(u) = 0$ for some $P_m \in \mathcal{P}_m$, where the number m is minimal. This means that the functions $u(x)$, $u(x+1), \ldots, u(x+m)$ are linearly dependent and the functions $u(x), \ldots, u(x+m-1)$ are linearly independent. Then, by Theorem 2.1,

$$D[u(x), \ldots, u(x+m)] = 0, \qquad \forall x \in Z_+.$$

We now show that there exists an $x \geq 1$ such that $D[u(x), \ldots, u(x+m-1)] \neq 0$. Suppose that, conversely, $D[u(x), \ldots, u(x+m-1)] = 0$ for $x \geq 1$. Again, by Theorem 2.1, we obtain that for $x \geq 1$ the functions $u(x), \ldots, u(x+m-1)$ are linearly dependent, i.e., for a certain k, $0 \leq k \leq m-1$

$$u(x+m-1-k) + c_1 u(x+m-1-k-1) + \ldots + c_{m-k-1} u(x) = 0$$

for all $x \geq 1$. Making the change of variables $x - 1 - k = y$, we obtain

$$u(y+m) + c_1 u(y+m-1) + \ldots + c_{m-k-1} u(y+k+1) = 0, \quad \geq 0.$$

Chapter 1. Inverse problems for difference operators

Subtracting this equation from the equation

$$P_m u \equiv u(y+m) + a_1 u(y+m-1) + \ldots + a_m u(y) = 0$$

we have

$$(a_1 - c_1)u(y+m-1) + \ldots + (a_{m-k-1} - c_{m-k-1})u(y+k+1)$$
$$+ a_{m-k} u(y-k) + \ldots + a_m u(y) = 0.$$

If $a_1 \neq c_1$, then u satisfies the difference equation of order $m-1$, which contradicts the minimality of m. Therefore, $a_1 = c_1$. Analogously, $a_2 = c_2$, ..., $a_{m-k-1} = c_{m-k-1}$ and $a_{m-k}u(y+k)+\ldots+a_m u(y) = 0$, i.e., the function u satisfies the equation of order not greater then $k < m$, which contradicts the minimality of m. The theorem is proved. $\square$

Theorem 2.3 provides a solution to Problem 2.3 in the case of $n = \infty$. Now we shall study Problem 2.3 for finite n. Set

$$Z_+^N = \{0, 1, \ldots, N\} = Z_+ \cap \{x \leq N\}.$$

Theorem 2.4. Let the function $u : Z_+ \to \mathbb{C}$ satisfy the following conditions for a certain m such that $2m \leq n - 1$:

1. For all $x \in Z_+^{n-2m-1}$ $D[u(x), u(x+1), \ldots, u(x+m)] = 0$.

2. There exists an $x \in Z_+^{n-2m+1}$, $x \geq 1$, such that

$$D[u(x), \ldots, u(x+m-1)] \neq 0.$$

Then there exists a unique operator $P_m \in \mathcal{P}_m$ and a unique function $\hat{u} : Z_+ \to \mathbb{C}$ such that

$$P_m \hat{u} = 0, \qquad \hat{u}(x) = u(x), \qquad x \in Z_+^{n-1}. \tag{2.16}$$

Moreover, conditions 1 and 2 are necessary for (2.16) to hold.

Proof. The necessity of conditions 1 and 2 for the validity of equalities (2.16) follows immediately from Theorem 2.3. To establish the sufficiency, we set, as before,

$$D_m(x) = D[u(x), \ldots, u(x+m)].$$

By assumption, $D_m(x) = 0$, $x \in Z_+^{n-2m-1}$. Therefore, from identity (2.12) we have

$$D_{m-1}^2(x+1) = D_{m-1}(x)D_{m-1}(x-2), \quad x \in Z_+^{n-2m-1}.$$

From these relations, arguing as in the proof of Theorem 2.3, we show that
$$D_{m-1}(x) \neq 0, \quad x \in Z_+^{n-2m+1}.$$

Extend the definition of u for $x = n$ by the condition

$$D_m(u - 2m) = \begin{vmatrix} u(n-2m) & u(n-2m-1) & \cdot & u(n-2m+m) \\ \cdot & \cdot & \cdot & \cdot \\ u(n-m) & u(n-m-1) & \cdot & u(n) \end{vmatrix} = 0.$$

Expanding this determinant with respect to the last row we uniquely determine $u(n)$ since the algebraic complement of the element $u(n)$ is equal to $D_{m-1}(n-2m) \neq 0$. So, for the extended function u we have

$$D_m(x) = 0, \quad x \in Z_+^{n-2m}$$

$$D_{m-1}(x) \neq 0, \quad x \in Z_+^{n-2m+2}$$

(the last relation is established due to equality (2.13) for $x = n - 2m + 2$). Defining $u(x)$ for the points $x = n+1, n+2, \ldots$ in a similar way, we thus construct a function $\hat{u}$ for which

$$D[\hat{u}(x), \ldots, \hat{u}(x+m)] = 0, \quad x = Z_+,$$

$$D[\hat{u}(x), \ldots, \hat{u}(x+m-1)] \neq 0, \quad x \in Z_+,$$

where, by construction, the restriction of $\hat{u}$ to Z_+^{n-1} coincides with u. From Theorem 2.3 it follows that there exists a unique operator $P_m \in \mathcal{P}_m$ such that $P_m \hat{u} = 0$. The theorem is proved. □

Exercise 2.1. Prove Theorems 2.1 and 2.2.

Exercise 2.2. Construct an example which shows that the condition $x \geq 1$ in Theorem 2.3 cannot be replaced by the condition $x \geq 0$.

Exercise 2.3. Establish formula (2.12) (Directions: see Pólya and Szegö (1964)).

Chapter 1. Inverse problems for difference operators

Exercise 2.4. Suppose A is the Frobenius matrix from Exercise 1.4 and $f \in \mathbb{C}^N$ is an arbitrary nonzero vector. Prove that the first m numbers in the infinite sequence consisting of the row-vectors $f, Af, A^2 f, \ldots$ uniquely determine the matrix A (i.e., the numbers $N, a_1, a_2, \ldots, a_N$) if $m \geq 2N + 1$.

Exercise 2.5. Find the law of formation of the sequences

$$0, 1, 2, 4, 12, 28, 68, \ldots, \qquad 1, 0, 2, 6, 10, 18, 50, \ldots,$$

$$1, 0, 2, 3, -1, -5, 0, \ldots.$$

1.3. THE PROBLEMS OF DETERMINING A DIFFERENCE OPERATOR IN NONSTATIONARY STATEMENT

In this section we state the inverse problems of the following type. Suppose A is a linear bounded operator acting in a Hilbert space X. We consider the difference Cauchy problem

$$u^{m+1} = Au^m, \qquad u^0 = f \in X, \tag{3.1}$$

whose solution, evidently, is given by the formula $u^m = A^m f$. Given a projection of the solution

$$g^m = Pu^m = PA^m f, \qquad m = 1, \ldots, M, \tag{3.2}$$

$$P \in \mathcal{L}(X), \qquad P^2 = P$$

it is required to reconstruct the operator A.

The same problem is considered for the differential equation

$$\frac{du}{dt} = Au, \qquad t \in [0, T], \tag{3.3}$$

$$u(0) = f, \tag{3.4}$$

$$g^m = Pu(t_m), \tag{3.5}$$

where $t_m = m\tau$, $m = 1, \ldots, M$, $\tau M = T$. The set X is assumed to be finite-dimensional. The operators A are difference operators. Inverse problems (3.1)–(3.2), (3.3)–(3.5) arise in discretization of the corresponding inverse problems for differential equations. A typical example is as follows.

Example 3.1 [The problem of determining the coefficient of a parabolic equation]. Suppose $u(x,t)$ is a smooth solution of the boundary value problem

$$u_t = u_{xx} + a(x)u, \qquad x \in [0,1], \quad t \in [0,T], \qquad (3.6)$$
$$u(0,t) = u(1,t) = 0, \qquad t \in [0,T], \qquad (3.7)$$
$$u(x,0) = f(x). \qquad (3.8)$$

We assume that the coefficient $a(x)$ is unknown. We need to find it using the additional conditions imposed on the function $u(x,t)$ on the boundary:

$$u(0,t) = g(t), \qquad t \in [0,T]. \qquad (3.9)$$

The uniqueness of the solution of this inverse problem under certain conditions imposed on the function f will be proved in Chapter 3. For the approximate solution of this problem we introduce a uniform mesh with spacing h with respect to the variable x:

$$x_j = (j+1)h, \qquad j = -1, 0, 1, \ldots, N$$

and a uniform mesh with spacing τ with respect to time t:

$$t_m = m\tau, \qquad m = 0, \ldots, M, \quad M\tau = T.$$

We set

$$u_j^m = u(x_j, t_m), \qquad a_j = a(x_j), \qquad f_j = f(x_j), \qquad g^m = g(t_m)$$

and approximate the problem (3.6)–(3.9) by the explicit difference scheme

$$\frac{u_j^{m+1} - u_j^m}{\tau} = \frac{u_{j+1}^m - 2u_j^m + u_{j-1}^m}{h^2} + a_j u_j^m,$$

$$u_{-1}^m = 0, \qquad u_N^m = 0, \qquad u_j^0 = f_j, \qquad u_0^m = g^m.$$

Introducing the column vectors

$$u^m = (u_0^m, u_1^m, \ldots, u_{N-1}^m)^T, \qquad f = (f_0, f_1, \ldots, f_{N-1})^T$$

(the symbol T denotes transposition) and the matrix

$$L = \frac{1}{h^2} \begin{pmatrix} a_0 h^2 - 2 & 1 & 0 & \cdot & 0 \\ 1 & a_1 h^2 - 2 & 1 & \cdot & 0 \\ \cdot & \cdot & \cdot & \cdot & \cdot \\ 0 & 0 & 0 & 1 & a_{N-1} h^2 - 2 \end{pmatrix}$$

we rewrite the difference scheme in the compact form
$$u^{m+1} - u^m = \tau L u^m, \qquad u^0 = f.$$

Now, setting $A = E + \tau L$, where E is the unit matrix, $Pu^m = u_0^m$, we pass to the inverse problem (3.1), (3.2) formulated above. Evidently, for determination of the operator A it suffices to find the numbers a_j from diag A, $j = 0, \ldots, N-1$. The uniqueness of the solution to this problem for $M = N$ follows from Theorem 3.1 established below. Another way of solution of the inverse problem (3.6)–(3.9) can be proposed. In this case we approximate the differential operator only with respect to x. Setting
$$u(t) = (u_0(t), u_1(t), \ldots, u_{N-1}(t))^T$$
we get the inverse problem (3.3)–(3.5), where $A = E + \tau L$, $g^m = u_0(m\tau)$.

We start the investigation of the inverse problem (3.1)–(3.2) with the case where the difference operator A acting in the space $l_2(0, N-1)$ of complex-valued sequences $u = (u_0, u_1, \ldots, u_{N-1})$ is given by the formula
$$(Au)_j = a_{j-1} u_{j-1} + b_j u_j + c_j u_{j+1}, \qquad j = 0, \ldots, N-1, \qquad (3.10)$$
where we set
$$u_{-1} = u_N = 0. \qquad (3.11)$$

Analogously to Example 3.1, we assume the complex numbers a_j and c_j to be given and the numbers b_j to be unknown. We need to determine them by the inverse problem data $g^m = (A_m f)_0$, $m = 1, \ldots, N$. Thus, the operator P from (3.2) is the operator of projection onto the first component of the vector u.

Theorem 3.1. *Suppose z_j, c_j and a vector $f \in l_2(0, N-1)$ are such that $f_j \neq 0$ for all $j = 0, \ldots, N-1$ and $c_j \neq 0$ for all $j = 0, \ldots, N-2$. Then for any sequence g^m, $m = 1, \ldots, N$, there exists a unique operator A of the form (3.10), (3.11) such that $(A^m f)_0 = g^m$ for all $m = 1, \ldots, N$.*

Proof. We write the equation $u^{m+1} = Au^m$ for the jth component:
$$u_j^{m+1} = a_{j-1} u_{j-1}^m + b_j u_j^m + c_j u_{j+1}^m. \qquad (3.12)$$

For $m = 0$, $j = 0$ we have $g^1 = b_0 f_0 + c_0 f_1$ (for the vector f, as for the vectors u^m, we assume that boundary conditions (3.11) hold). Then
$$b_0 = \frac{g^1 - c_0 f_1}{f_0}.$$

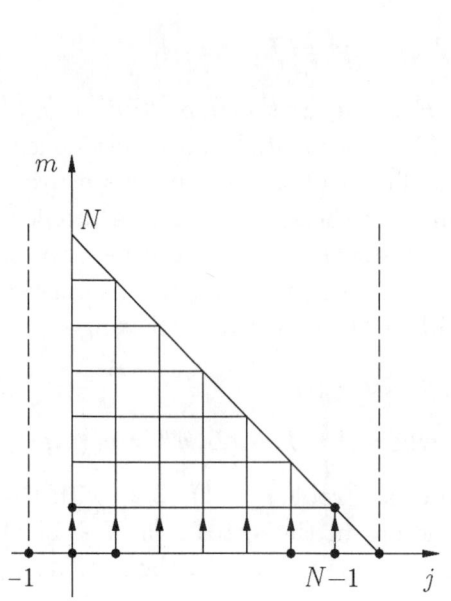 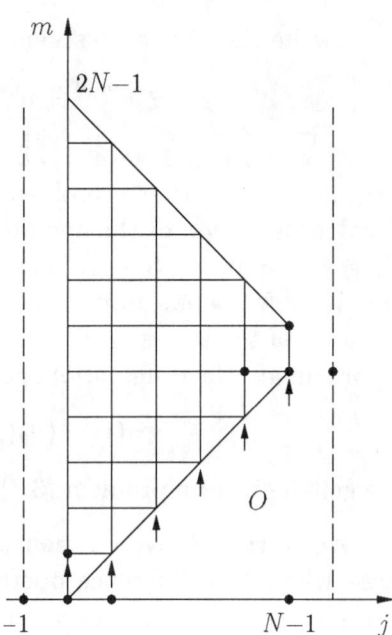

Figure 1 Figure 2

Now, setting successively $m = 1, 2, \ldots, N-1$, $j = 0$, by the condition $c_0 \neq 0$ and formula (3.12) we uniquely determine u_1^m, $m = 1, \ldots, N-1$. Further, we repeat the process for $j = 1$: for $m = 0$, $j = 1$ we obtain b_1, for $m = 1, \ldots, N-2$, $j = 1$ we obtain u_2^m, and so on. As a result, the numbers b_j are obtained recurrently by the formulas

$$b_j = \frac{u_j^1 - a_{j-1}f_{j-1} - c_j f_{j+1}}{f_j}, \qquad u_{j+1}^m = \frac{u_j^{m+1} - a_{j-1}u_{j-1}^m - b_j u_j^m}{c_j},$$

$$u_0^m = g^m, \qquad j = 0, \ldots, N-1, \quad m = 0, \ldots, N-j-1.$$

The process of successive reconstruction of b_j and u_{j+1}^m is shown in Figure 1, where by large dots we denote the initial and final positions of the pattern of the difference scheme (3.12) (i.e., of the nodes (j, m) which enter formula (3.12)). The arrows show the direction of its motion along the vertical lines $j = 0, 1, \ldots, N-2$. The dotted lines denote the boundary conditions. The theorem is proved. □

In the next theorem, a special choice of the initial vector f allows us to determine not only b_j but also a_j if $a_j \neq 0$.

Chapter 1. Inverse problems for difference operators

Theorem 3.2. *Suppose $f = \delta_0 = (1, 0, \ldots, 0)$ and the numbers $c_j \ne 0$ are given for $j = 0, \ldots, N-2$. Then the sequence $g^m = (A^m f)_0$, $m = 1, \ldots, 2N-1$, uniquely determines an operator A of the form (3.10), (3.11), where $a_j \ne 0$, $j = 0, \ldots, N-2$.*

Proof. First, we note that since $f_j = 0$ for $j > 0$, taking into account the Cauchy data $u^0 = f$, from equation (3.12) we have

$$u_j^m = 0, \qquad j > m \qquad (3.13)$$

for any coefficients a_j, b_j, c_j. To determine b_0 we set $j = 0$, $m = 0$ in (3.12). Then, since $u_0^0 = f_0 = 1$, $u_0^1 = g^1$, and $u_{-1}^0 = 0$, we obtain $b_0 = g^1$. Now, setting $j = 0$, $m = 1, 2, \ldots, 2N - 2$ we find the numbers

$$u_1^m = \frac{g^{m+1} - b_0 g^m}{c_0}.$$

For $j = 1$, $m = 0$, from (3.12) and (3.13) we have $a_0 = u_1^1$. Thereupon the process repeats: for $j = 1$, $m = 1$ we find b_1, then, setting $m = 2, 3, \ldots, 2N - 3$, we obtain u_2^m; for $j = 2$, $m = 1$ we find a_1, and so on. As a result, the desired coefficients b_j and a_j are defined by the recurrent formulas

$$b_j = \frac{u_j^{j+1} - a_{j-1} u_{j-1}^j}{u_j^j}, \qquad j = 0, \ldots, N-1, \qquad (3.14)$$

$$u_{j+1}^m = \frac{u_j^{m+1} - a_{j-1} u_{j-1}^m - b_j u_j^m}{c_j}, \qquad m = j+1, j+2, \ldots, 2N-2-j, \quad (3.15)$$

$$a_j = \frac{u_{j+1}^{j+1}}{u_j^j}, \qquad j = 0, \ldots, N-2, \qquad (3.16)$$

$$u_0^0 = 1, \qquad u_0^m = g^m, \qquad m = 1, \ldots, 2N-1.$$

The fact that $u_j^j \ne 0$, $j = 1, \ldots, N-1$, follows from formula (3.16) and the condition $a_j \ne 0$, $j = 0, \ldots, N-2$. The process of successive reconstruction of b_j, a_{j+1}^n, a_j is shown in Figure 2. As before, large dots denote the initial and final positions of the pattern of the difference scheme, and arrows denote the directions of its motion along the lines $j = 0, 1, \ldots, N-1$. The theorem is proved. □

Remark 3.1. Expressing the condition $u_j^j \neq 0$ in terms of the initial data g^m we may prove the existence theorem.

Now, let the numbers a_j and c_j in formula (3.10) be complex conjugate and
$$a_j = \bar{c}_j \neq 0, \qquad j = 0, \ldots, N-2. \tag{3.17}$$

Theorem 3.3. *Suppose that condition (3.17) holds and $\arg a_j$ (the arguments of a_j) are given, $j = 0, \ldots, N-2$. Then the sequence*
$$g_m = (A^m f)_0, \qquad m = 1, \ldots, 2N-1,$$
where $f = \delta_0 = (1, 0, \ldots, 0)$, determines the operator A uniquely.

Proof. Setting $m = j + 1$ in formula (3.12), we have
$$u_j^{j+2} = a_{j-1} u_{j-1}^{j+1} + b_j u_j^{j+1} + \bar{a}_j u_{j+1}^{j+1}. \tag{3.18}$$

From formula (3.16) it follows that
$$\bar{a}_j u_{j+1}^{j+1} = \bar{a}_j a_j u_j^j.$$

Therefore, formula (3.18) uniquely determines the absolute values of a_j:
$$|a_j| = \left\{ \frac{u_j^{j+2} - a_{j-1} u_{j-1}^{j+1} - b_j u_j^{j+1}}{u_j^j} \right\}^{1/2}, \qquad j = 0, \ldots, N-1. \tag{3.19}$$

In particular, $|a_0| = \sqrt{g^2 - b_0 g^1}$. As in Theorem 3.2, the numbers b_j and u_{j+1}^m, $m \geq j+1$, are defined by formulas (3.14) and (3.15) (in (3.15) we take $c_j = \bar{a}_j$). In this case, in contrast to Theorem 3.2, we change the order of determination of the parameters b_j, a_j, u_{j-1}^m. First, by formula (3.14) we find b_j. Then we find a_{j+1}^m from equality (3.15). The theorem is proved. □

At first glance, it may seem that if in the assumptions of Theorem 3.1 the sequence g^m is known not up to $M = N$ but up to $2N - 1$, then, as in Theorems 3.2 and 3.3, we obtain the numbers b_j and a_j simultaneously. However, this is not so. As a counterexample we may take the matrices A_1 and A_2 from Example 1.1 such that $(A_1^m f)_0 = (A_2^m f)_0$ for all integers $m \geq 0$, where $f = (1, 1)^T$. The cause of the nonuniqueness is that we have a considerable nonlinear "mixing" of the sought parameters a_j and b_j in the case of the vector f with all nonzero components. Therefore, to reconstruct several

Chapter 1. Inverse problems for difference operators

diagonals of A we need to have the same number of vectors f. We shall prove the corresponding theorem in the case of a self-adjoint band matrix A with $2l+1$ nonzero diagonals. It is convenient to consider such operator in the difference form

$$(Au)(x) = \sum_{k=-l}^{l} a_k(x) u(x+k), \qquad x = 0, \ldots, N-2, \qquad (3.20)$$

$$u(-l) = u(-l+1) = \ldots = u(-1) = 0, \qquad (3.21)$$
$$u(N) = u(N+1) = \ldots = u(N+l-1) = 0. \qquad (3.22)$$

Here the number l is such that $1 \leq l \leq N-1$ and the self-adjointness condition has the form

$$a_{-k} = \bar{a}_k(x-k). \qquad (3.23)$$

Taking into account this condition, to determine the operator A it suffices to find $(l+1)$ coefficients $a_0(x), a_1(x), \ldots, a_l(x)$. Therefore, we assume that for any $x = -l, \ldots, N+l-1$ $u(x)$ is a vector function of dimension $l+1$

$$u : Z_{-l}^{N+l-1} \to \mathbb{C}^{l+1}$$

which satisfies the boundary conditions (3.21), (3.22). Suppose u^m is a solution of the difference Cauchy problem (3.1), where $f \in l_2(Z_{-l}^{N+l-1}; \mathbb{C}^{l+1})$ and the operator A is defined by formulas (3.20)–(3.22).

Setting $M = N+1-l$, for the operator P we take the projector on the first l values $0, 1, \ldots, l-1$ of the vector $u^m(x)$ for $m \leq M-1$ and $Pu^m = u^m(0)$ for $m = M$.

Thus, in this example the operator P depends on m:

$$Pu^m = \begin{cases} u^m|_{Z_0^{l-1}}, & m \leq M-1, \\ u^m(0), & m = M. \end{cases}$$

For $j = 0, \ldots, N-1$ we set $p(j) = \min(l, N-j-1)$,

$$\mathcal{D}(j) = (f(j), f(j+1), \ldots, f(j+p)).$$

We shall write the vectors $f(j)$ as columns. Then $\mathcal{D}_p(j)$ will be a matrix of size $(l+1) \times (p+1)$.

Theorem 3.4. Suppose that for all $x = 0, \ldots, N-1$

$$\operatorname{rank} \mathcal{D}_p(x) = p+1 \qquad (3.24)$$

and $a_l(x) \neq 0$ for $0 \leq x \leq N-1-l$. Then the sequence $g^m = PA^m f$, $m = 1, \ldots, N+1-l$, uniquely determines the operator A.

Proof. Setting $x = 0$ and $m = 0$ in the equality

$$u^{m+1} = Au^m \qquad (3.25)$$

we have

$$u^1(0) = \sum_{k=0}^{l} a_k(0) f(k). \qquad (3.26)$$

By assumption, we have $u^1(0) = g^1(0)$ and rank $\mathcal{D}_p(0) = l+1$. Therefore, we uniquely determine the coefficients $a_k(0)$, $k = 0, \ldots, l$, from the linear system (3.26). Since $a_l(0) \neq 0$, for $x = 0$, $m = 1, \ldots, N - l$ from equation (3.25) we derive

$$u^m(l) = \left[u^{m+1}(0) - \sum_{k=0}^{l-1} a_k(0) u^m(k) \right] a_l^{-1}(0).$$

Suppose that at the $(j-1)$th step the following numbers are already determined $a_k(x)$, $x = 0, 1, \ldots, j-1$, $x+k \leq N-1$ and $u^m(l+j-1)$, $m = 1, \ldots, N-1-j+1$. Setting $x = j$ and $m = 0$ in (3.25), we have

$$u^1(j) = \sum_{k=-l}^{l} a_k(j) f(j+k).$$

Taking into account the boundary conditions $f(j+k) = 0$ for $j+k \geq N$, by the definition of the number $p(j)$, we need to find the coefficients $a_0(j)$, $a_1(j)$, ..., $a_p(j)$. The coefficients $a_{-k}(j)$ with negative indices, by formula (3.23), are expressed in terms of $a_k(x)$ with $x < j$ and therefore they are known. The uniqueness of determination of the coefficients $a_k(j)$, $k = 0, \ldots, p(j)$ follows from condition (3.24). If $j \leq N - l - 1$, then we derive $u^m(l+j)$ by the formula

$$u^m(l+j) = \left[u^{m+1}(j) - \sum_{k=-l}^{l-1} a_k(j) u^m(j+k) \right] a_l^{-1}(j)$$

Otherwise, $u^m(l+j) = 0$ since boundary conditions (3.22) hold. The theorem is proved. □

The generalization of Theorems 3.1–3.4 and similar theorems to the case of partial difference operators is evident. For this reason we shall restrict ourselves to the formulation of multidimensional analogue of Theorem 3.1.

Chapter 1. Inverse problems for difference operators

First, we shall give the definition of a partial difference operator in a convenient form.

To this end, at first, we must take a certain finite subset J of the integer lattice Z^n, called *a pattern*. For an arbitrary integer domain $\Omega \subset Z^n$, we set

$$\partial \Omega = \{y \in Z^n \setminus \Omega \mid y = x + \alpha, \quad x \in \Omega, \quad \alpha \in J\}.$$

The set $\partial \Omega$ is called *the boundary of the domain Ω relative to the pattern J*. For the functions of discrete argument $u : Z^n \to \mathbb{C}$ we define the multidimensional shift operator $H^\alpha = H_1^{\alpha_1} \ldots H_n^{\alpha_n}$ by the formula

$$H^\alpha u(x) = u(x_1 + \alpha_1, x_2 + \alpha_2, \ldots, x_n + \alpha_n),$$

where $x, \alpha \in Z^n$. Here H_j is the operator of the shift by 1 with respect to the variable x_j. We define an operator A by the formula

$$(Au)(x) = \sum_{\alpha \in J} a_\alpha(x) H^\alpha u(x), \tag{3.27}$$

where $a_\alpha : Z^n \to \mathbb{C}$ are given functions called the coefficients of the operator.

In order to define the restriction of the operator A to the domain Ω correctly, in the general case (for example, if $a_\alpha(x) \neq 0$ and $0 \in J$) it is necessary to know the values of the function u on the set $\Omega \cup \partial\Omega$, i.e., to know the boundary conditions

$$u|_{\partial\Omega} = \varphi. \tag{3.28}$$

Then the formulas (3.27) and (3.28) correctly define $Au(x)$ for $x \in \Omega$.

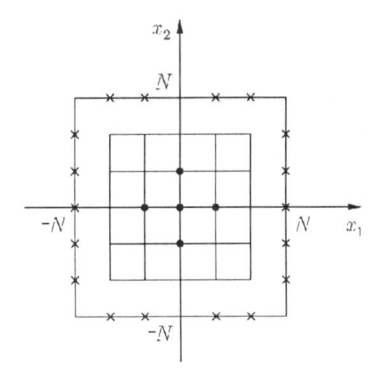

Figure 3

Example 3.2. Suppose that $n = 2$ and $J = \{\alpha \in Z^2 \mid |\alpha_1| + |\alpha_2| \leq 1\}$, $\Omega = \{x \in Z^2 \mid |x_j| \leq N - 1\}$. In Figure 3 the boundary $\partial\Omega$ is designated by crosses and the pattern J by large dots. If we set

$$a_\alpha(x) = \begin{cases} h^{-2}, & \alpha \neq 0, \\ -2h^{-2} + b(x), & \alpha = 0, \end{cases} \quad \alpha \in J$$

then the operator A is the difference analogue of the operator

$$\Delta u(x) + b(x)u, \qquad x \in \mathbb{R}^2$$

on the square mesh with spacing $h > 0$.

For the simplicity of formulation, we consider the case where

$$\Omega = \{x \in Z^n \mid |x_j| \le M_j, \quad j = 1, \ldots, n-1, \quad 0 \le x_n \le N-1\},$$

$$J = \{\alpha \in Z^N \mid |\alpha_1| + |\alpha_2| + \ldots + |\alpha_n| \le 1\}.$$

Denote $\Gamma = \omega \cup \{x \mid x_n = 0\}$. Then we may prove the following theorem analogously to Theorem 3.1.

Theorem 3.5. *Suppose that for $\alpha \ne 0$ the coefficients of an operator A and a function $f : \Omega \to \mathbb{C}$ are given. Let $f(x) \ne 0$ and $a_\alpha(x) \ne 0$ for $\alpha = (0, 0, \ldots, 0, 1)$ and for all $x \in \Omega$. Then, for a given sequence of functions $g^m : \Gamma \to \mathbb{C}$, $m = 1, \ldots, N$, there exists a unique operator A of the form (3.27), (3.28) such that*

$$u^{m+1} = Au^m, \qquad u^0 = f, \quad x \in \Omega,$$

$$u^m|_{\partial\Omega} = \varphi, \quad u^m|_\Gamma = g^m.$$

Theorems 3.1–3.5 may be easily reformulated in spectral terms if the operator A is self-adjoint. In this case, in particular, we obtain some refined results of Section 1. For example, we consider the operator A defined by formulas (3.10), (3.11) in the case of $a_j = \bar{c}_j$, $b_j \in \mathbb{R}$. Then $A = A^*$ and from the definition of the measure $d\rho(\lambda)$ constructed in Section 1 it follows that for any vector $f \in l_2(0, N-1)$ we have

$$f = \int \langle f, p(\lambda) \rangle p(\lambda) \, d\sigma(\lambda) = \int p(\lambda) \, d\rho(\lambda).$$

Since $Ap(\lambda) = \lambda p(\lambda)$ for $\lambda \in \operatorname{Sp} A$,

$$A^m f = \int \lambda^m p(\lambda) \, d\rho(\lambda).$$

In particular, taking into account that $p_0(\lambda) = 1$, we obtain

$$g^m = \int \lambda^m \, d\rho(\lambda). \tag{3.29}$$

Thus, formulas (3.29) and Theorem 3.1 imply the following theorem.

Chapter 1. Inverse problems for difference operators

Theorem 3.6. *Suppose that numbers $a_j = \bar{c}_j \neq 0$, $j = 0,\ldots,N-2$ are given, and a vector $f \in l_2(0, N-1)$ is such that $f_i \neq 0$ for all $j = 0,\ldots,N-1$. Then $\operatorname{diag} A$ is uniquely determined by the following N moments of the measure $d\rho(\lambda)$:*

$$\int \lambda^m \, d\rho(\lambda), \qquad m = 1, \ldots, N.$$

If $f = \delta_0 = (1, 0, \ldots, 0)$, then $\langle f, p(\lambda) \rangle = p_0(\lambda) = 1$, $d\rho(\lambda) = d\sigma(\lambda)$,

$$g^m = \int \lambda^m \, d\sigma(\lambda). \tag{3.30}$$

Therefore, we obtain the following statement from Theorem 3.3.

Theorem 3.7. *Suppose $a_j = \bar{c}_j \neq 0$ and the arguments of the numbers a_j, $j = 0, \ldots, N-2$, are given. Then the spectral function $\sigma(\lambda)$ uniquely determines the operator A.*

In the general case, if under the hypotheses of Theorem 3.3 the number N (the order of the matrix A) is unknown and, for simplicity, the numbers a_j are assumed to be positive, then the following theorem holds.

Theorem 3.8. *Suppose $0 \notin \operatorname{Sp} A$. Then the sequence $g^m = (A^m f)_0$, $m = 1, \ldots, M - 1$, where $f = \delta_0$ and $M \geq 2N - 1$, uniquely determines the operator A, i.e., the numbers N, $a_j > 0$, $j = 0, \ldots, N - 2$, b_j, $j = 0, \ldots, N - 1$.*

Proof. Since

$$d\sigma(\lambda) = \sum_{k=1}^{N} \varkappa_k \delta(\lambda - \lambda_k),$$

where $\varkappa_k = \|p(\lambda_k)\|^{-2} \neq 0$ and δ is the generalized Dirac function, from formula (3.30) we have

$$g^m = \sum_{k=1}^{N} \lambda_k^m \varkappa_k, \qquad m = 1, \ldots, M - 1.$$

Observing that $g^0 = 1$, $\lambda_k \neq 0$, by Theorem 2.4, we can uniquely determine the numbers $N, \varkappa_k, \lambda_k$, i.e., find the measure $d\delta(\lambda)$. Therefore, by Theorem 3.7, we find the whole operator A. The theorem is proved. □

Now, we study the uniqueness of the solution of the inverse problem (3.3), (3.5), assuming for simplicity that the operator A defined by formulas (3.10) and (3.11) is self-adjoint and $a_j > 0$.

Theorem 3.9. *Suppose $u(t)$ is a solution of the Cauchy problem*

$$u_t = Au, \qquad u(0) = \delta_0.$$

Then the sequence $g^m = u_0(m\tau)$, $m = 1, \ldots, M-1$ for $M \geq 2N+1$ uniquely determines the operator A, i.e., the numbers N, a_j, b_j. If the number N is given, then uniqueness holds for $M = 2N$.

Proof. Since $u(t) = \exp(At)$, we have

$$g^m = \sum_{k=1}^{N} \exp(m\tau\lambda_k)\varkappa_k.$$

Setting $\xi_k = \exp(\tau\lambda_k)$, by Theorem 2.4, we find the numbers N, ξ_k, $\varkappa_k$ and, consequently, λ_k, i.e., the spectral function $\sigma(\lambda)$. If the number N is given, in order to determine λ_k and $\varkappa_k$ it suffices to know the numbers g^m with m from 0 up to $2N-1$. The reference to Theorem 3.7 finishes the proof. □

The following theorem is proved in a similar way.

Theorem 3.10. *Suppose $u(t)$ is a solution to the Cauchy problem*

$$u_t = Au, \qquad u(0) = f,$$

where $f \in l_2(0, N-1)$, $f_i \neq 0$ for all $j = 0, \ldots, N-1$, and N is given. Then diag A is uniquely determined by the sequence $g^m = u(m\tau)$, $m = 1, \ldots, 2N$.

In Theorems 3.1–3.5 and Example 1.1 we study inverse problems for explicit difference schemes. It was essential because for implicit difference schemes, in the general case, we have no uniqueness of the solution. To show this, we consider the following difference scheme, which is implicit for $\sigma \neq 0$:

$$\frac{u^{m+1} - u^m}{\tau} = A(\sigma u^{m+1} + (1-\sigma)u^m), \qquad u^0 = f, \qquad (3.31)$$

where A is a self-adjoint three-diagonal matrix defined by formulas (3.10), (3.11) ($a_j = \bar{c}_j \neq 0$) and σ is an arbitrary complex parameter. We show that for $\sigma \neq 0$ the analogue of Theorem 3.1 fails for this equation. Indeed, if we have the sequence $g^m = u_0^m$, $m = 1, \ldots, M$, then, setting $j = 0$, $m = 0, 1, \ldots, M-1$ in (3.31) for $\sigma \neq 0$, we obtain a linear system with M equations and $M+1$ unknowns b_0, u_1^m, $m = 1, \ldots, M$. Thus, we have no

Chapter 1. Inverse problems for difference operators

recurrent algorithm for determining b_j and u_{j+1}^m. Of course, this does not mean that we cannot define b_j uniquely for any M. We shall consider an example where $M = N$ and the analogue of Theorem 3.1 fails. Suppose $N = 2$, $\sigma = 1$, $\tau = 1$, $f = (1,1)^T$. Then the operators

$$A = \begin{pmatrix} 0 & -1 \\ -1 & 0 \end{pmatrix}, \qquad \overline{A} = \begin{pmatrix} -2 & -1 \\ -1 & 2 \end{pmatrix}$$

generate the same sequence g^m, $m = 1, 2$, since for $u^0 = \overline{u}^0 = f$ we have

$$u^1 = (1/2, 1/2)^T, \qquad u^2 = (1, 1/2)^T,$$

$$\overline{u}^1 = (1/2, -1/2)^T, \qquad \overline{u}^2 = (0, 1/2)^T$$

and therefore $u_0^m = \overline{u}_0^m$, $m = 0, 1, 2$.

Now we show that, in the general case, if we make the sequence g^m longer and impose certain restrictions on σ, then the solution to the inverse problem is unique.

Theorem 3.11. *Suppose that the numbers $(\sigma\tau)^{-1}$, $1 + (\sigma\tau)^{-1}$, $-\tau^{-1}(1-\sigma)^{-1}$ are not contained in the spectrum of the operator A. If $f_j \neq 0$ for $j = 0, \ldots, N-1$, then the sequence $g^m = u_0^m$, $m = 1, \ldots, 2N$, uniquely determines $\mathrm{diag}\, A$.*

Proof. Solving equation (3.31) for u^{m+1}, we obtain

$$u^{m+1} = Su^m, \qquad u^0 = f,$$

where the transition operator is

$$S = S(A) = (E - \sigma\tau A)^{-1}(E + \tau(1-\sigma)A).$$

Hence, we obtain $u^m = S^m f$ and, in particular,

$$g^m = \sum_{k=1}^{N} \varkappa_k S^m(\lambda_k), \qquad m = 0, \ldots, 2N, \qquad (3.32)$$

$$\varkappa_k = \langle f, p(\lambda_k) \rangle \, \|p(\lambda_k)\|^{-2}.$$

By Theorem 2.4, from system (3.32) we uniquely determine the nonzero numbers $\varkappa_k$ and the corresponding numbers $S(\lambda_k)$, which are nonzero by the assumption of the theorem.

In view of the condition $1 + (\sigma\tau)^{-1} \notin \operatorname{Sp} A$, we have $1 - \sigma + \sigma S_k \neq 0$ and therefore the formula

$$\lambda_k = \frac{s_k - 1}{\tau(1 - \sigma + \sigma s_k)}$$

uniquely determines the spectrum of the operator A and the measure

$$d\rho(\lambda) = \sum_k \varkappa_k \delta(\lambda - \lambda_k).$$

Application of Theorem 3.6 completes the proof. $\square$

Remark 3.2. The assumptions of Theorem 3.11 hold automatically if $\operatorname{Im} \sigma \neq 0$.

The example considered above shows that explicit (with respect to time) equations have certain advantages. However, since the implicit equations are more preferable when solving direct problems as far as stability is concerned, we may use an implicit equation for computing the initial data g^m and an explicit one for solving the inverse problem. The recurrent algorithms of Theorems 3.1–3.5 show that the main sources of error accumulation appear mostly when computing u_{j+1}^m, i.e., when solving the difference Cauchy problem for the variable j. In a typical case, for instance, in the conditions of Example 3.1, the continuous analogue of this problem is an ill-posed Cauchy problem. The principles of construction of difference equations for such ill-posed boundary value problems and the conditions of their stability and convergence are discussed in Chapter 6. Note that the stability of these difference schemes is interpreted in a certain generalized sense because the classical theory of stability is not applicable here since the limit (differential) problem is ill-posed.

Exercise 3.1. Suppose that the operator A is defined by formula (3.10) and the boundary conditions (3.11) are replaced by the following ones:

$$u_{-1} = 0, \qquad u_N + h n_{N-1} = 0,$$

where $h \in \mathbb{R}$, $h \neq 0$. How do the formulations of Theorems 3.1–3.3 change if the number h is known? Show that if b_{N-1} and c_{N-1} are given, we can uniquely reconstruct the operator A and the number h under the assumptions of Theorems 3.1–3.3.

Exercise 3.2. Suppose that an operator A which acts in the space $l_2(0, N-1)$ is defined by the formula

$$(Au)_j = a_{j-1}u_{j-1} + (b_j - a_j - a_{j-1})u_j + a_j u_{j+1},$$

$$u_{-1} = 0, \qquad u_N = 0, \qquad a_{-1} = 1.$$

Such operator arises in the difference approximation of the differential operator $\partial/\partial x(a(x)\partial u/\partial x) + b(x)u$.

Prove the analogues of Theorems 3.1, 3.3, and 3.4 for the operator A.

1.4. REMARKS AND REFERENCES

Chapter 1 is introductory. Its main purpose is to motivate the statements of inverse problems for differential equations. At the same time, as the example of the difference scheme (3.31) shows, in the general case, however natural discretization may seem from the viewpoint of solution of direct problems, in the case of inverse problems it can lead to the loss of the uniqueness of solutions.

To Section 1. Theorems 1.1–1.4 are proved by the author in Bukhgeim (1986a). The problem of reconstruction of an infinite Jacobi matrix from the spectral function was investigated in details by Berezanskii (1965). In the case of finite real matrices this problem was considered by Atkinson (1964). These results are connected with the research of Stieltjes (1938) on continued fractions. The problem of reconstruction of a Jacobi matrix by the spectral function has an interesting application in the integration of some nonlinear difference equations arising in physics in the explicit form (see Berezanskii (1985)).

To Section 2. This section contains the classical theory taken from Gel'fond (1967). It is interpreted from the point of view of inverse problems. The main result of the section is Theorem 2.4, which was proved in Proni (1975).

To Section 3. The main results of this section, probably, have not been published in such version. Theorems 3.1, 3.4, and 3.5 are difference analogues of the statements of inverse problems that were first considered in Bukhgeim and Yakhno (1976).

Generally speaking, there are many works devoted to inverse problems for difference equations. The corresponding references and results can be found in the monographs of Berezanskii (1965), Atkinson (1964), Shadon and Sabatier (1980), and in Zakharov and Suz'ko (1985). Some numerical aspects of solution of inverse problems by finite-difference methods are discussed in Kunetz (1963), Alekseev (1967), and in Alekseev and Dobrinskii (1975).

Chapter 2.

A priori estimates and the uniqueness of solutions of integro-differential equations with operator coefficients

2.1. ESTIMATES OF THE CARLEMAN TYPE AND THEIR CONNECTION WITH THE UNIQUENESS OF SOLUTIONS OF INVERSE PROBLEMS

Suppose that $\Omega \subset \mathbb{R}^2$ is an open set of real variables x, t and P is the differential Schrödinger operator with potential $a(x,t) \in L_\infty(\Omega)$:

$$Pu = \mathrm{i}\partial_t u + \partial^2 u + a(x,t)u. \tag{1.1}$$

Here $\partial_t = \partial/\partial t$, $\partial = \partial/\partial x$, $\mathrm{i}^2 = -1$. For a real function $\varphi \in C^\infty(\overline{\Omega})$ and a number $s \geq 0$ we introduce the set of weight L_2-norms

$$\|u\|_s^2 = \int e^{2s\varphi(x,t)} |u(x,t)|^2 \, dx\, dt, \tag{1.2}$$

where we put $\|u_0\| = \|u\|$ for brevity. Our next goal is to prove the a priori estimate

$$s^3 \|u\|_s^2 + s \|\partial u\|_s^2 \leq c \|Pu\|_s^2, \tag{1.3}$$

which must hold for all $u \in C_0^\infty(\Omega)$ and $s \geq s_0$ with some constant $c > 0$ independent of s and u. First, we set forth the known general method of establishing estimates similar to (1.3), which we shall often use in this chapter.

Inequality (1.3) is proved by means estimating $\|Pu\|_s$ from below. Observing that
$$\|Pu\|_s = \|e^{s\varphi} P e^{-s\varphi} e^{s\varphi} u\|$$
it is convenient to pass from the operator P to the similar operator $L = e^{s\varphi} P e^{-s\varphi}$ and introduce the new function $v = e^{s\varphi} u$ instead of u. Then
$$\|Pu\|_s = \|Lv\|. \tag{1.4}$$

Further, to estimate $\|Lv\|$ we represent the operator L as the sum of the symmetric term L_+ and skew-symmetric term L_-, $L_\pm = (L \pm L^*)/2$, where the operator L^* is formally conjugate to L with respect to the scalar product $(\cdot, \cdot)$ in $L_2(\Omega)$. It is determined from the equality
$$(Lv, w) = (v, L^* w), \qquad v, w \in C_0^\infty(\Omega).$$

Since $L = L_+ + L_-$ and
$$\|Lv\|^2 = \|L_+ v\|^2 + \|L_- v\|^2 + 2\operatorname{Re}(L_+ v, L_- v),$$
taking into account the equalities $L_+^* = L_+$, $L_-^* = -L_-$ we have
$$2\operatorname{Re}(L_+ v, L_- v) = (L_+ v, L_- v) + (L_- v, L_+ v)$$
$$= -(L_- L_+ v, v) + (L_+ L_- v, v) = ([L_+, L_-]v, v).$$

Here $[L_+, L_-] = L_+ L_- - L_- L_+$ is the commutator of the operators L_+ and L_-. Thus,
$$\|Lv\|^2 = \|L_+ v\|^2 + \|L_- v\|^2 + ([L_+, L_-]v, v). \tag{1.5}$$

The aim of the following considerations is to estimate the quadratic form $([L_+, L_-]v, v)$ from below by means of integration by parts. Thereupon, we return from the function v to the function u. One more general remark is that it suffices to establish an estimate of the form (1.3) for the operator P' which may differ from P by terms inside the norm signs in the left-hand side of (1.3), i.e., in this case,
$$\|(P' - P)u\|_s \leq c(\|u\|_s + \|\partial u\|_s). \tag{1.6}$$

Indeed, if
$$s^3\|u\|_s^2 + s\|\partial u\|_s^2 \le c\|P'u\|_s^2 \tag{1.7}$$
for all $s \ge s_0$, then, since $P' = P + P' - P$, from the triangle inequality and (1.6) it follows that
$$\|P'u\|_s^2 \le (\|Pu\|_s + \|(P - P')u\|_s)^2 \le 2\|Pu\|_s^2 + 4c^2(\|u\|_s^2 + \|\partial u\|_s^2).$$
Substituting this estimate into (1.7), we obtain
$$s^3(1 - 4c^3 s^{-3})\|u\|_s^2 + s(1 - 4c^3 s^{-1})\|\partial u\|_s^2 \le 2c\|Pu\|_s^2.$$
Increasing the number s_0, if necessary, so that
$$\min(1 - 4c^3 s_0^{-3}, 1 - 4c^3 s_0^{-1}) \ge 1/2,$$
we obtain estimate (1.3) with the constant $4c$ instead of c. Applied to the operator P of the form (1.1), this consideration means that it suffices to prove estimate (1.3) for $P = i\partial_t + \partial^2$. In this case we find the operators $L_\pm$ and their commutator. We have
$$L = i\partial_t + \partial^2 - 2s\varphi_x\partial + s^2\varphi_x^2 - s\varphi_{xx} - is\varphi_t.$$
Since $(\varphi_x\partial)^* = -\partial\varphi_x = -\varphi_x\partial - \varphi_{xx}$, we get
$$L^* = i\partial_t + \partial^2 + 2s\varphi_x\partial + s^2\varphi_x^2 + s\varphi_{xx} + is\varphi_t.$$
Consequently, we have
$$L_+ = i\partial_t + \partial^2 + s^2\varphi_x^2, \qquad L_- = -2s\varphi_x\partial - s\varphi_{xx} - is\varphi_t$$
which implies
$$[L_+, L_-] = a_2\partial^2 + a_1\partial + a_0$$
with certain coefficients $a_j(x, t, s)$. To find them, we note the following evident formulas which hold for any smooth functions $\alpha(x,t)$ and $\beta(x,t)$:
$$[\beta\partial^2, \alpha\partial] = (2\beta\alpha_x - \alpha\beta_x)\partial^2 + \beta\alpha_{xx}\partial,$$
$$[\beta\partial^2, \alpha] = 2\beta\alpha_x\partial + \beta\alpha_{xx}, \qquad [\beta\partial, \alpha\partial] = (\beta\alpha_x - \alpha\beta_x)\partial,$$
$$[\beta\partial, \alpha] = \beta\alpha_x, \qquad [\partial_t, \alpha\partial] = \alpha_t\partial.$$
Using these commutation formulas and the linearity of the commutator in both arguments, we find
$$a_2 = -4s\varphi_{xx}, \qquad a_1 = -4s(\varphi_{xx} + i\varphi_t)_x = O(s),$$

$$a_0 = 4s^3\varphi_x^2\varphi_{xx} - i(s\varphi_{xx} + is\varphi_t)_t - (s\varphi_{xx} + is\varphi_t)_{xx}.$$

Integrating by parts in the scalar product $(a_2\partial^2 v, v)$, we get

$$(a_2\partial^2 v, v) = 4s(\varphi_{xx}\partial v, \partial v) + 4s(\varphi_{xxx}\partial v, v). \tag{1.8}$$

From the form of a_0 it follows that

$$(a_0 v, v) = 4s^3(\varphi_x^2\varphi_{xx}v, v) + O(s\|v\|^2). \tag{1.9}$$

The scalar products $(a_1\partial v, v)$ and $s(\varphi_{xxx}\partial v, v)$ are estimated by means of the Cauchy-Schwarz-Bunyakovskii inequality and the α-inequality:

$$2xy \le \alpha x^2 + \alpha^{-1}y^2, \qquad x, y \ge 0, \quad \alpha > 0,$$

$$|(a_1\partial v, v)| + s|(\varphi_{xxx}\partial v, v)| \le 2cs\|\partial v\|\,\|v\| \le \alpha\|\partial v\|^2 + \alpha^{-1}c^2 s^2\|v\|^2.$$

Hence, taking into account (1.8) and (1.9), on condition that $\alpha = 1$ and $|\varphi_x| \ge 1$, $\varphi_{xx} \ge 1$, we have

$$([L_+, L_-]v, v) \ge s^3(4 - c^2 s^{-1} - cs^2)\|v\|^2 + s(4 - s^{-1}\|\partial v\|^2)$$

with some constant. Choosing the number s_0 so that the multipliers in parentheses for $s \ge s_0$ are greater or equal to unity and using (1.4) and (1.5), we obtain

$$s^3\|v\|^2 + s\|\partial v\|^2 \le ([L_+, L_-]v, v) \le \|Pu\|_s^2.$$

Recalling that $v = \exp(s\varphi)u$, $\partial v = \exp(s\varphi)(\partial u + s\varphi_x u)$, we obtain

$$\|v\| = \|u\|_s,$$

$$s\|\partial v\|^2 = s\|\partial u + s\varphi_x u\|_s^2 \ge s\|\partial u\|_s^2 - 2s^2 c\|\partial u\|_s\|u\|_s$$
$$\ge s\|\partial u\|_s^2 - \alpha s\|\partial u\|_s^2 - s^3 c^2 \alpha^{-1}\|u\|_s^2.$$

We have used the α-inequality in the last estimate. Setting $\alpha = 1/2$, multiplying the last inequality by $\varepsilon \in (0, 1)$, and adding $s^3\|u\|^2$, we obtain

$$s^3(1 - 2\varepsilon c^2)\|u\|_s^2 + (s\varepsilon/2)\|\partial u\|_s^2 \le s^3\|v\|^2 + s\varepsilon\|\partial v\|^2 \le \|Pu\|_s^2.$$

Choosing ε so small that $1 - 2\varepsilon c^2 \ge \varepsilon/2$, we obtain estimate (1.3) with constant $c = 2/\varepsilon$. So, we have proved the following statement.

Chapter 2. A priori estimates and the uniqueness of solutions

Proposition 1.1. Suppose that

$$|\varphi_x| \geq 1, \qquad \varphi_{xx} \geq 1. \qquad (1.10)$$

for all $x, t \in \overline{\Omega}$. Then for $a \in L_\infty(\Omega)$ there exist numbers s_0 and $c > 0$ such that estimate (1.3) holds for all $s \geq s_0$ and $u \in C_0^\infty(\Omega)$.

Now, we obtain an estimate for the operator

$$P = i\partial_t + b(x)\partial^2 + d(x,t)\partial - a(x,t), \qquad (1.11)$$

where $a, d \in L_\infty(\Omega)$, $b \in C^3(\overline{\Omega})$. As in the proof of Proposition 1.1, the estimate for the operator P will be established if we prove it for the operator $P = i\partial_t + \partial b \partial$, which differs from (1.11) only by lower terms. In this case we have

$$L = i\partial_t + b\partial^2 + (b_x - 2s\varphi_x b)\partial + (s^2\varphi_x^2 - s\varphi_{xx})b - s\varphi_x b_x - is\varphi_t,$$
$$L^* = i\partial_t + b\partial^2 + b_x\partial + 2s\varphi_x b\partial + 2s(\varphi_x b)_x$$
$$\quad + (s^2\varphi_x^2 - s\varphi_{xx})b - s\varphi_x b_x + is\varphi_t.$$

Hence, we obtain

$$L_+ = i\partial_t + b\partial^2 + b_x\partial + s^2\varphi_x^2 b, \qquad L_- = -2s\varphi_x b\partial - s((\varphi_x b)_x + i\varphi_t).$$

As before, $[L_+, L_-] = a_2\partial^2 + a_1\partial + a_0$, where, by the commutation formulas, we have

$$a_2 = -4s\varphi_{xx}b^2 - s\varphi_x(b^2)_x,$$
$$a_1 = O(s) \qquad \text{for} \quad b \in C^2(\overline{\Omega}),$$
$$a_0 = 2s^3\varphi_x b(\varphi_x^2 b)_x + O(s) \qquad \text{for} \quad b \in C^3(\overline{\Omega}).$$

Hence, for

$$\varphi_x b(\varphi_x^2 b)_x \geq \delta > 0 \qquad (1.12)$$

we have the estimate

$$(a_0 v, v) \geq 2s^3\delta\|v\|^2 - cs\|v\|^2 \geq s^3\delta\|v\|^2$$

with some constant $c > 0$ if $s \geq s_0$ and the number s_0 is sufficiently large. The form $(a_2\partial^2 v, v)$ is transformed, as before, by means of integration by parts:

$$(a_2\partial^2 v, v) = -(a_2\partial v, \partial v) - (a_{2x}\partial v, v).$$

The absolute values of the quadratic forms $(a_{2x}\partial v, v)$ and $(a_1\partial v, v)$ are estimated from above by the value $\|\partial v\|^2 + cs^2\|v\|^2$ using the α-inequality. Thus, if we assume that

$$-a_2/s = 4\varphi_{xx}b^2 + \varphi_x 2bb_x \geq \delta > 0, \tag{1.13}$$

then

$$([L_+, L_-]v, v) \geq s^3(\delta - cs^{-1})\|v\|^2 + s(\delta - s^{-1})\|\partial v\|^2 \geq s^3\delta'\|v\|^2 + s\delta'\|\partial v\|^2$$

for $\delta' < \delta$ and $s \geq s_0$, where s_0 is sufficiently large. Recalling that

$$\|Pu\|_s^2 \geq ([L_+, L_-]v, v)$$

and returning from the function v to the function u as in the proof of Proposition 1.1, we obtain the following statement.

Proposition 1.2. Let $b \in C^3(\overline{\Omega})$, $a, d \in L_\infty(\Omega)$, and

$$s\varphi_{xx}b^2 + \varphi_x b_x b \geq \varepsilon, \qquad \varphi_x^2 \geq \varepsilon \tag{1.14}$$

for all $(x, t) \in \overline{\Omega}$. Then there exist numbers s_0 and $c > 0$ such that for an operator of the form (1.11) estimate (1.3) holds.

To prove this proposition, it suffices to note that conditions (1.12) and (1.13) follow from the assumption (1.14).

Now, we consider a simple example in order to show the connection between estimate (1.3) and the uniqueness of the solution to the inverse problem of determining one of the coefficients of the operator P of the form (1.11), assuming that it depends only on the spatial variable x.

So, let $u(x, t)$ be a smooth solution of the following boundary value problem for the Schrödinger equation:

$$iu_t + u_{xx} - a(x)u = 0, \qquad x \geq 0, \quad t \in \mathbb{R}, \tag{1.15}$$

$$u_x(0, t) = 0, \qquad u(x, 0) = f(x). \tag{1.16}$$

In the general case, the potential $a(x)$ and the initial data $f(x)$ may assume complex values. Given the trace of $u(x, t)$ for $x = 0$,

$$u(0, t) = g(t), \qquad t \in [-T, T], \tag{1.17}$$

and the functions f and g, it is required to reconstruct the potential $a(x)$. We show that the uniqueness of the solution of this inverse problem follows

Chapter 2. A priori estimates and the uniqueness of solutions

from the uniqueness of the solution of the Cauchy problem with data given at $x = 0$ for the integro-differential equation, whose principal part is the Schrödinger operator. For this purpose, we assume that the problem (1.15)–(1.17) has two solutions u_1, a_1, u_2, a_2 which satisfy the same boundary conditions. Subtract equation (1.15) written for u_2, a_2 from equation (1.15) written for u_1, a_1. Denoting $u_1 - u_2 = u$ and $a_1 - a_2 = a$, we obtain

$$iu_t + u_{xx} - a_1 u = a(x)u_2(x,t), \tag{1.18}$$

$$u(x,0) = 0, \tag{1.19}$$

$$u(0,t) = 0, \quad u_x(0,t) = 0, \quad t \in [-T,T]. \tag{1.20}$$

For brevity, set $P = i\partial_t + \partial^2 - a_1$, $h(x,t) = a(x)u_2(x,t)$. Then equation (1.18) takes the form

$$Pu = h. \tag{1.21}$$

Thus, it is required to find the solution u and the right-hand side h of the differential equation (1.21) by the given traces of the solution u and u_x on the lines $x = 0$ and $t = 0$. Since there are no boundary conditions for $h(x,t)$, it is natural to exclude it from equation (1.21). This can be done observing that the function h satisfies the following ordinary differential equation of the first order

$$u_2(x,t)h_t - (\partial_t u_2)h = 0.$$

Assume that $u_2(x,t) \neq 0$. Then this equation may be written in terms of the operator Q:

$$Qh = (\partial_t - b(x,t))h = 0, \tag{1.22}$$

$$b(x,t) = (\partial_t u_2)/u_2. \tag{1.23}$$

Applying the operator Q to equation (1.21) and transposing it with the operator P, we obtain

$$PQu = [P,Q]u. \tag{1.24}$$

Equation (1.24) can be written as a system of equations if we introduce the new unknown function $v = Qu$:

$$Pv = [P,Q]u, \tag{1.25}$$

$$Qu = v. \tag{1.26}$$

This form is more convenient since the operator Q transforms the zero Cauchy data (1.20) for the function u into the zero Cauchy data for the function v. Indeed,

$$v(0,t) = (Qu)(0,t) = u_t(0,t) - b(0,t)u(0,t) = 0, \tag{1.27}$$

$$v_x(0,t) = u_{tx} - b_x u - b u_x|_{x=0} = 0, \qquad (1.28)$$

since $u_{xt}(0,t) = 0$, which is obtained by differentiating the second condition of (1.20) with respect to t. Thus, the system (1.25), (1.26) represents two Cauchy problems for the functions v and u with the Cauchy data at $x = 0$ and $t = 0$, respectively, connected by the commutator $[P,Q]$. Using the boundary condition $u(x,0) = 0$, the Cauchy problem for the equation $Qu = v$ can be solved explicitly:

$$Qu = u_2(x,t)\partial_t(u/u_2) = v \quad \Rightarrow$$

$$u(x,t) = Q^{-1}v = \int_0^t K(x,t,\tau)v(x,\tau)\,d\tau, \qquad (1.29)$$

where

$$K(x,t,\tau) = u_2(x,t)/u_2(x,\tau). \qquad (1.30)$$

Thus, we exclude the function u from the system (1.25), (1.27):

$$Pv = [P,Q]Q^{-1}v, \qquad (1.31)$$

$$v(0,t) = 0, \qquad v_x(0,t) = 0. \qquad (1.32)$$

Since the principal parts of the operators P and Q (i.e., the terms which contain the derivatives of maximal order) commute in our case, we have

$$[P,Q] = [i\partial_t + \partial^2 - a_1, \partial_t - b] = -[i\partial_t + \partial^2, b] - [a_1, \partial_t] = -ib_t - b_{xx} - 2b_x\partial - a_{1t}.$$

Taking into account that $a_{1t} = 0$, we have

$$[P,Q] = b_1\partial + b_0, \qquad (1.33)$$

where $b_1 = -2b_x$, $b_0 = -ib_t - b_{xx}$. Thus, equation (1.31) in detailed representation is the following integro-differential equation of the form

$$iv_t + v_{xx} = a_1 v + \int_0^t \{(b_1 K'_x + b_0 K)v(x,\tau) + b_1 K v_x(x,\tau)\}\,d\tau. \qquad (1.34)$$

To prove the uniqueness of the solution of the Cauchy problem (1.32), (1.34) we need the estimates of the integral operator

$$(\mathcal{J}u)(t) = \int_0^t u(\tau)\tau$$

in the weight norm

$$\|u\|_s^2 = \int_0^T e^{2s\varphi(t)}|u(t)|^2\,dt, \qquad s \geq 0.$$

These estimates are established by the following lemma.

Chapter 2. A priori estimates and the uniqueness of solutions

Lemma 1.1. Let $\varphi \in C^2[0,T]$ and $-\varphi'(t) \geq 0$ for all $t \in [0,T]$. Then
$$\|\mathcal{J}u\|_s^2 \leq T^2 \|u\|_s^2. \tag{1.35}$$
If, in addition, we suppose that $-\varphi''(t) \geq \mu > 0$ for all $t \in [0,T]$, then
$$s\|\mathcal{J}u\|_s^2 \leq c\|u\|_s^2, \qquad c = \pi/4\mu. \tag{1.36}$$

Proof. Let $v = e^{s\varphi}u$, $w = e^{s\varphi}\mathcal{J}u$. Then
$$w(t) = \int_0^t \exp(s(\varphi(t) - \varphi(\tau)))v(\tau)\,d\tau.$$
Since $\varphi(t)$ is an increasing function, $\varphi(t) - \varphi(\tau) \leq 0$ for $t \geq \tau$ and
$$\|w(t)\| \leq (\mathcal{J}|v|)(t).$$
As the norm of the integral operator $\mathcal{J}$ in the space $L_2(0,T)$ does not exceed T, we have $\|w\|^2 \leq T^2 \|v\|^2$. Hence, returning to the function u, we obtain estimate (1.35). Under the additional condition $-\varphi''(t) \geq \mu > 0$ the Taylor formula for $t \geq \tau$ yields
$$\varphi(t) - \varphi(\tau) \leq -\mu(t-\tau)^2/2.$$
Therefore,
$$|w(t)| \leq \int_0^t k(t-\tau)|v(\tau)|\,d\tau, \tag{1.37}$$
$$k(t) = \exp(-\mu s t^2/2).$$
The right-hand side of estimate (1.37) is the convolution $k * |v|$ of the function $|v|$ with the kernel k. From the Young inequality, it follows that
$$\|k * v\| \leq \|k\|_{L_1(0,T)} \|v\|.$$
In our case, we obtain
$$\|k\|_{L_1(0,T)} \leq \int_0^\infty \exp(-\mu s t^2)\,dt = (\mu s)^{-1/2} \int_0^\infty e^{x^2}\,dx = \sqrt{\frac{\pi}{2\mu s}}$$
and therefore
$$\|w\|^2 \leq (\pi/4\mu s)\|v\|^2.$$
Multiplying both sides of the inequality by s and returning to the function u, we obtain estimate (1.36). The lemma is proved. $\square$

Proposition 1.3. Let $\varphi \in C^2[-T,T]$,

$$\|u\|_s^2 = \int_{-T}^{T} e^{2s\varphi(t)} |u(t)|^2 \, dt, \qquad s \geq 0,$$

and for all $t \in [-T,T]$

$$-\varphi'(t) \operatorname{sign} t \geq 0 \tag{1.38}$$

where

$$\operatorname{sign} t = \begin{cases} 1, & t \geq 0, \\ -1, & t < 0. \end{cases}$$

Then estimate (1.35) holds for the operator $\mathcal{J}$. Under the additional assumption that $-\varphi''(t) \geq \mu$ for all $t \in [-T,T]$ estimate (1.36) is valid.

To prove this proposition, it suffices to expand the space $L_2(-T,T)$ into the direct sum $L_2(-T,0) \oplus L_2(0,T)$ and apply Lemma 1.1 to each of the spaces after making the change of variables $t \to -t$ in the case of $L_2(-T,T)$.

We now return to the proof of uniqueness for the Cauchy problem (1.32), (1.34). First, we note that (1.34) implies the inequality

$$|iv_t + v_{xx}| \leq c(|v(x,t)| + |(\mathcal{J}|v|)(x,t)| + |(\mathcal{J}|v_x|)(x,t)|). \tag{1.39}$$

As we shall see later, the major fact that allows us to apply estimates of Propositions 1.1 and 1.3 to the proof of uniqueness for the Cauchy problem is that the weight function decreases with increasing distance from a point of the space to the support of the Cauchy data (in our case, it is the segment $[-T,T]$ of the axis t). Besides, when deducing equation (1.34) and inequality (1.39), we used the fact that $u_2(x,t) \neq 0$ in the domain $\overline{\Omega}$. In terms of the initial data, this condition will hold if $f(x) \neq 0$ for all $x \in \mathbb{R}$. Then, taking for Ω a domain concentrated in a small neighbourhood of the segment $[0,\eta]$ of the x-axis we may guarantee (in view of continuity) that

$$u_2(x,t) \neq 0, \qquad \forall (x,t) \in \overline{\Omega}. \tag{1.40}$$

An example of such a domain is provided by the half of the ellipse interior

$$\Omega = \{x, t \in \mathbb{R} \mid \psi(x,t) > 0, \quad x > 0\}, \tag{1.41}$$

$$\psi(x,t) = 1 - (x/r)^2 - (t/b)^2$$

with sufficiently small semiaxis b.

Chapter 2. A priori estimates and the uniqueness of solutions

The function ψ decreases with increasing distance from the point (x, t) to the segment $[-T, T]$ of t-axis, on which the Cauchy data are given. However, we cannot take it as a weight function since it does not satisfy the conditions of Proposition 1.1. Indeed, $\psi_x = 0$ for $x = 0$ and, moreover, $\psi_{xx} < 0$. But its modification

$$\varphi(x, t) = \exp(\lambda\psi(x + \delta, t)) - 1 \tag{1.42}$$

with $\delta > 0$ and sufficiently large parameter $\lambda > 0$ satisfies all the conditions of Proposition 1.1. The domain

$$\Omega = \{x, t \in \mathbb{R} \mid \varphi(x, t) > 0, \quad x > 0\} \tag{1.43}$$

is transformed into the domain defined by inequality (1.41) as $\delta \to 0$.

Indeed, since for $(x, t) \in \overline{\Omega}$, $\delta \in (0, 1)$,

$$-\psi_x(x + \delta, t) = 2(x + \delta)/r^2 \geq 2\delta/r^2,$$

$$\psi_{xx} = -2/r^2, \qquad \psi(x + \delta, t) \geq 0,$$

for $\lambda \geq \lambda_0$, $\lambda_0 = r^2/\delta$, we have

$$-\varphi_x = -\lambda\psi_x \exp(\lambda\psi) \geq 2\delta\lambda/r^2 \geq 2,$$

$$\varphi_{xx} = \exp(\lambda\psi)(\lambda^2\psi_x^2 + \lambda\psi_{xx}) \geq (\lambda^2 4\delta^2/r^4 - 2\lambda/r^2) \geq \delta^{-1}(4 - 2) \geq 2.$$

Moreover, $\psi_t = -2t/b^2$ implies $-\varphi_t \operatorname{sign} t \geq 2\lambda|t|/b^2 \geq 0$ and therefore the first estimate of Proposition 1.3 holds. Also, we may satisfy the condition $-\varphi_{tt} \geq \mu > 0$ for $\lambda < 1/2$. However, we shall not use this condition here.

So, suppose that the function φ and the domain Ω are defined by formulas (1.42) and (1.43) with $\lambda \geq \lambda_0$ and the number b is so small that (1.40) holds. Then inequality (1.39) holds for $v = Qu$. Moreover, $v(0, t) = v_x(0, t) = 0$, $(0, t) \in \partial\Omega$. We set $\Omega_\varepsilon = \Omega \cap \{\varphi > \varepsilon\}$ and suppose that χ is a smooth function such that $0 \leq \chi \leq 1$ in Ω. In this case, we suppose that $\chi = 1$ for $(x, t) \in \Omega_\varepsilon$ and $\chi = 0$ for $(x, t) \in \Omega \setminus \Omega_{\varepsilon/2}$. Evidently, such a function exists. By continuity, the estimate (1.3) of Proposition 1.1 holds for any function $u \in C^2(\overline{\Omega})$ such that u and u_x vanish on the boundary $\partial\Omega$ of the

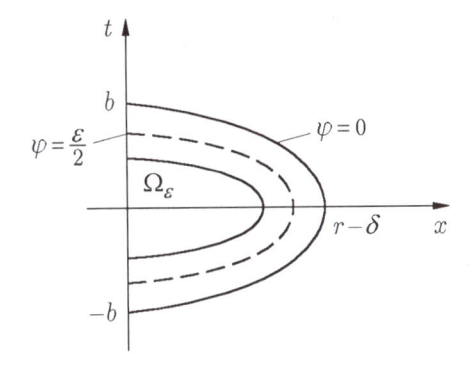

Figure 4

domain Ω. Therefore, for such a function we may take χv. Indeed, it vanishes together with $(\chi v)_x$ on the boundary $x = 0$ of the domain Ω because the Cauchy data are zero. In the neighbourhood of the boundary $\varphi = 0$ this function is equal to zero by the construction of the function χ (see Figure 4).

Substituting the function χv into estimate (1.3), for $s \geq \max(1, s_0)$ we have

$$s \int_{\Omega_\varepsilon} e^{2s\varphi}(|v(x,t)|^2 + |v_x(x,t)|^2) \, dx \, dt$$
$$\leq s \int_\Omega e^{2s\varphi}(|\chi v|^2 + |(\chi v)_x|^2) \, dx \, dt \leq c \int_\Omega e^{2s\varphi}|\mathrm{i}(\chi v)_t + (\chi v)_{xx}|^2 \, dx \, dt$$
$$= c \int_{\Omega_\varepsilon} e^{2s\varphi}|\mathrm{i} v_t + v_{xx}|^2 \, dx \, dt + c \int_{\Omega \setminus \Omega_\varepsilon} e^{2s\varphi}|\mathrm{i}(\chi v)_t + (\chi v)_{xx}|^2 \, dx \, dt.$$

(1.44)

Estimate (1.39) and the Cauchy-Schwarz-Bunyakovskii inequality for the sums yield

$$\int_{\Omega_\varepsilon} e^{2s\varphi}|\mathrm{i} v_t + v_{xx}|^2 \, dx \, dt \leq 3c^2 \int_{\Omega_\varepsilon} e^{2s\varphi}(|v|^2 + |\mathcal{J}|v||^2 + |\mathcal{J}|v_x||^2) \, dx \, dt.$$

(1.45)

Applying the first estimate of Proposition 1.3 to the integrals of $|\mathcal{J}|v||^2$ and $|\mathcal{J}|v_x||^2$ with respect to t, we can estimate the right-hand side of inequality (1.45) from above by the quantity

$$c \int_{\Omega_\varepsilon} e^{2s\varphi}(|v|^2 + |v_x|^2) \, dx \, dt$$

with some other constant c. Consequently,

$$\int_{\Omega_\varepsilon} e^{2s\varphi}|\mathrm{i} v_t + v_{xx}|^2 \, dx \, dt \leq c \int_{\Omega_\varepsilon} e^{2s\varphi}(|v|^2 + |v_x|^2) \, dx \, dt.$$

Combining this estimate with inequality (1.44), we obtain

$$(s - c^2) \int_{\Omega_\varepsilon} e^{2s\varphi}(|v|^2 + |v_x|^2) \, dx \, dt \leq c \int_{\Omega \setminus \Omega_\varepsilon} e^{2s\varphi}|\mathrm{i}(\chi v)_t + (\chi v)_{xx}|^2 \, dx \, dt.$$

(1.46)

For $s \geq s_0$, if the number s_0 is chosen so that $1 - c^2/s_0 \geq 1/2$, we have $s - c^2 \geq s/2$. Besides, the minimum of the function φ in the domain Ω_ε coincides with its maximum in $\Omega \setminus \Omega_\varepsilon$ and is equal to ε by construction

Chapter 2. A priori estimates and the uniqueness of solutions

(see Figure 4). Taking the quantity $\min \exp(2s\varphi) = \exp(2s\varepsilon)$ out from the integral over Ω_ε and taking the maximum of the weight $\exp(2s\varepsilon)$ out from the integral over $\Omega \setminus \Omega_\varepsilon$, we, evidently, do not violate equality (1.46). After dividing by the multiplier $\exp(2s\varepsilon)$ and taking into account that $s - c^2 \geq s/2$ for $s \geq s_0$, we obtain

$$s \int_{\Omega_\varepsilon} (|v|^2 + |v_x|^2) \, dx \, dt \leq c \int_{\Omega \setminus \Omega_\varepsilon} |i(\chi v_t) + (\chi v)_{xx}|^2 \, dx \, dt.$$

Dividing both parts of the inequality by s and passing to the limit as $s \to \infty$, we have

$$\int_{\Omega_\varepsilon} (|v|^2 + |v_x|^2) \, dx \, dt = 0.$$

Hence, $v = 0$ in Ω_ε. Passing to the limit as $\varepsilon \to 0$, we find that $v = 0$ in Ω. Taking into account (1.29), we have $u = Q^{-1}v$ and, consequently, $u = 0$. Therefore, $Pu = h = u_2(x,t)a(x) = 0$. By condition (1.40), $u_2 \neq 0$ for all $(x,t) \in \overline{\Omega}$ and thus $a(x) = 0$. This means that $a_1(x) = a_2(x)$ for all $x \in [0, r - \delta]$. Since r and s are arbitrary, $a_1(x) = a_2(x)$ for all $x \geq 0$. So, we have proved the following statement.

Theorem 1.1. *Let $u(x,t)$ be a solution of the boundary value problem (1.15), (1.16) which is three times continuously differentiable in the half-plane $x \geq 0$. Suppose also that $f(x) = u(x,0) \neq 0$ for all $x \geq 0$ and the coefficient $a \in C[0, \infty)$. Then the potential $a(x)$ is uniquely determined by the function $g(t) = u(0,t)$, $t \in (-T, T)$.*

A similar procedure can be used to prove the uniqueness of determination of the potential of the Schrödinger equation

$$iu_t + \Delta u - a(x)u = 0$$

in the multidimensional case ($x \in \mathbb{R}^n$). In this case it suffices to establish estimate (1.3) for the operator $P = i\partial_t + \partial^2 - A(x)$ with operator potential $A(x) = -\Delta' + a(x)$, where $\partial = \partial/\partial x_n$, Δ' is the Laplace operator with respect to the variables $x' = (x_1, \ldots, x_{n-1})$. The estimates for differential operators of two variables x, t with operator coefficients are considered in Section 2 of this chapter. The uniqueness theorems based on these estimates are proved in Section 3 of Chapter 3.

Exercise 1.1. Prove estimate (1.3) for the heat conduction operator $P = \partial_t - \partial^2$.

Exercise 1.2. Prove estimate (1.3) for the operator $P = \partial_t - b(x)\partial^2 - d(x)\partial - a(x)$ and apply it to obtain the uniqueness theorem for the problem of finding one of the coefficients (assuming the others to be known) of the following boundary value problem:

$$Pu = 0, \qquad x \geq 0, \; t \in \mathbb{R}$$

$$u(x,t) = f(x), \qquad u_x(0,t) = 0, \qquad u(0,t) = g(t), \qquad t \in (-T, T).$$

2.2. ESTIMATES FOR THE SCHRÖDINGER EQUATION WITH OPERATOR COEFFICIENTS

Suppose that H is a complex Hilbert space with norm $|u|$ and scalar product $\langle u, v \rangle$, Ω is an open set of variables $(x,t) \in \mathbb{R}^2$. Denote by $L_2(\Omega; H)$ the Hilbert space of functions $u: \Omega \to H$ with norm

$$\|u\|^2 = \int_\Omega |u(x,t)|^2 \, dx \, dt \tag{2.1}$$

and scalar product

$$(u, v) = \int_\Omega \langle u(x,t), v(x,t) \rangle \, dx \, dt.$$

As in the previous section, along with the norm (2.1) we consider the following set of equivalent norms $L_2(\Omega; H)$:

$$\|u\|_s^2 = \int_\Omega e^{2s\varphi(x,t)} |u(x,t)|^2 \, dx \, dt.$$

Here $s \geq 0$ is a number, φ is a real function of class $C^\infty(\overline{\Omega})$. Suppose V is a linear manifold in H, where a linear operator

$$\Lambda: V \to V \times V \times \ldots \times V = V^n, \qquad n \geq 1,$$

is defined so that

$$|\Lambda u|^2 = \sum_{j=1}^n |(\Lambda u)_j|^2 \geq |u|.$$

Here $(\Lambda u)_j = \Lambda_j u$ is the jth coordinate of the vector Λu. By means of this operator we shall determine the order of linear operators $A(x): V \to V$ that may depend on the parameter x. We denote by $\mathcal{L}_m^k$ the class of the operators $A(x): V \to V$ that are k times strongly continuously differentiable

Chapter 2. A priori estimates and the uniqueness of solutions

with respect to x in their domain of definition, V, and such that for all $v \in V$ the estimate

$$\sum_{p=0}^{k} |A^{(p)}(x)v| \leq c|\Lambda^m v| \qquad (2.2)$$

holds with some constant $c = c_A$ independent of x. Here

$$\Lambda^m u = (\Lambda_1^m u, \ldots, \Lambda_n^m u).$$

Thus, we may say that $\mathcal{L}_m^k$ is the class of k times continuously differentiable operators of order not higher than m (with respect to the operator Λ). In particular, these operators are bounded in H if $m = 0$. We define a differential operator

$$P = i\partial_t + \partial B(x)\partial + D(x)\partial - A(x) \qquad (2.3)$$

with operator coefficients A, B, and D such that

$$A \in \mathcal{L}_2^1, \qquad B \in \mathcal{L}_0^3, \qquad D \in \mathcal{L}_1^3. \qquad (2.4)$$

We assume that the operators A and B are symmetric and D is skew-symmetric on V, i.e.,

$$\langle Au, v \rangle = \langle u, Av \rangle, \qquad \langle Bu, v \rangle = \langle u, Bv \rangle, \qquad \langle Du, v \rangle = -\langle u, Dv \rangle$$

for all $u, v \in V$.

We denote by $C_0^\infty(\Omega; V)$ the space of infinitely differentiable functions with values in V and with compact support in Ω. Our goal is to establish the *a priori* estimate

$$s^3 \|u\|_s^2 + s(\|\partial u\|_s^2 + \|\Lambda u\|_s^2) \leq c\|Pu\|_s^2 \qquad (2.5)$$

which holds for all $s \geq s_0$ and $u \in C_0^\infty(\Omega; V)$.

As in Section 1, we begin with the simplest case: $B(x) \equiv E$ (the unit operator) and $D(x) = 0$.

Theorem 2.1. Suppose $A \in \mathcal{L}_2^1$ and for all $v \in V$

$$\langle A'(x), v, v \rangle \geq c_0 |\Lambda v|^2 - c_1 |v|^2 \qquad (2.6)$$

with some constants $c_0, c_1 > 0$. If

$$-\varphi_x \geq 1, \qquad \varphi_{xx} \geq 1 \qquad (2.7)$$

then there exist numbers $c, s_0 > 0$ such that for any $u \in C_0^\infty(\Omega; V)$ the estimate (2.5) holds for the operator $P = i\partial_t + \partial^2 - A(x)$.

Proof. Following the general procedure of proving the estimates of the form (2.5) presented in Section 1, we need to find the operator $L = e^{s\varphi} P e^{-s\varphi}$ and the operators

$$L_\pm = (L \pm L^*)/2,$$

where L^* is formally conjugate to L and satisfies the identity

$$(Lv, w) = (v, L^*w), \qquad \forall v, w \in C_0^\infty(\Omega; V).$$

We have

$$L = i\partial_t + \partial^2 - 2s\varphi_x \partial + (s^2\varphi_x^2 - s\varphi_{xx}) - A - is\varphi_t,$$
$$L^* = i\partial_t + \partial^2 + 2s\varphi_x \partial + 2s\varphi_{xx} + (s^2\varphi_x^2 - s\varphi_{xx}) - A + is\varphi_t,$$

$$L_+ = i\partial_t + \partial^2 + s^2\varphi_x^2 - A, \qquad L_- = -2s\varphi_x \partial - s\varphi_{xx} - is\varphi_t.$$

Hence,

$$[L_+, L_-] = A_2 \partial^2 + A_1 \partial + A_0$$

with some operator coefficients $A_j(x, t, s)$. A simple calculation shows that

$$A_2 = -4s\varphi_{xx}, \qquad A_1 = -4s(\varphi_{xx} + i\varphi_t)_x = O(s),$$

$$A_0 = 4s^3 \varphi_x^2 \varphi_{xx} - is(\varphi_{xx} + i\varphi_t)_t - s(\varphi_{xx} + i\varphi_t)_{xx} - 2s\varphi_x A'(x).$$

In what follows, for simplicity we omit the symbol of the unit operator in the terms like φ_{xx}. The only difference from the scalar case considered in Proposition 1.1 is the term $-2s\varphi_x A'(x)$ in the expression for the coefficient A_0. For this term, using estimate (2.6) and the first condition from (2.7), we have

$$(-2s\varphi_x A'v, v) = -2s \int_\Omega \varphi_x \langle A'v, v \rangle \, dx \, dt$$

$$\geq 2s \int_\Omega \langle A'v, v \rangle \, dx \, dt \geq 2sc_0 \|Av\|^2 - 2sc_1 \|v\|^2. \quad (2.8)$$

The second term in the right-hand side of estimate (2.8) is majorized for large s by the term $s^3 \|v\|^2$, which appears in the quadratic form $([L_+, L_-]v, v)$ as we know from the proof of Proposition 1.1. Therefore, combining the estimate for $([L_+, L_-]v, v)$ from the proof of Proposition 1.1 with estimate (2.8) for $u = e^{-s\varphi}v$, we have

$$\|Pu\|_s^2 \geq ([L_+, L_-]v, v) \geq s^3 \|v\|^2 + s\|v_x\|^2 + 2sc_0 \|Av\|^2$$

Chapter 2. A priori estimates and the uniqueness of solutions

if $s \geq s_0$, where s_0 is chosen to be sufficiently large. Returning from the function v to the function u in the right-hand side of the inequality, as in the proof of Proposition 1.1, we obtain estimate (2.5). The theorem is proved.
□

For the general case, the basic condition which guarantees the estimate (2.5) is expressed in terms of the following operator

$$M = -s\varphi_{xx}D^2 - s\varphi_x\{DD_x - [A, D] + 2BA_x\} + 1/2\{[A, D_x] - DD_{xx}\}. \quad (2.9)$$

Theorem 2.2. *Suppose that condition (2.4) holds, $[A, B], [D, D_x] \in \mathcal{L}_1^0$, $[B, D_x], [B, D] \in \mathcal{L}_0^1$ and for all $v \in V$ and $s \geq s_0$, where s_0 is sufficiently large, the following inequalities hold:*

$$\langle Mv, v\rangle \geq sc_0|\Lambda v|^2 - sc_1|v|^2, \qquad c_0 > 0, \quad (2.10)$$

$$\langle B^2 v, v\rangle \geq \mu|v|^2, \qquad \mu > 0. \quad (2.11)$$

If $\varphi_x^2 \geq 1$ and for a sufficiently large number C_2 depending on c_0, c_1, and μ

$$\varphi_{xx} - c_2(|\varphi_x + 1|) \geq 1 \quad (2.12)$$

then the estimate (2.5) holds for all $u \in C_0^\infty(\Omega; V)$ and $s \geq s_0$.

Proof. Since

$$e^{s\varphi}\partial_t e^{-s\varphi} = \partial_t - s\varphi_t, \qquad e^{s\varphi}D\partial e^{-s\varphi} = D\partial - s\varphi_x D,$$

$$e^{s\varphi}\partial B\partial e^{-s\varphi} = B\partial^2 + (B_x - 2s\varphi_x B)\partial + (s^2\varphi_x^2 - s\varphi_{xx})B - s\varphi_x B_x,$$

the operator $L = e^{s\varphi}Pe^{-s\varphi}$ has the form

$$L = i\partial_t + B\partial^2 + (B_x - 2s\varphi_x B + D)\partial + C_+ + C_-,$$

where

$$C_+ = (s^2\varphi_x^2 - s\varphi_{xx})B - s\varphi_x B_x - A, \qquad C_- = -is\varphi_t - s\varphi_x D.$$

Since $(C_\pm)^* = \pm C_\pm$, we have

$$L^* = i\partial_t + B\partial^2 + B_x\partial + 2s\varphi_x B\partial + 2s(\varphi_x B)_x + D_x + D\partial + C_+ - C_-$$

Therefore, taking into account that

$$C_+ + s(\varphi_x B)_x = s^2\varphi_x^2 B - A,$$

we obtain

$$L_+ = i\partial_t + B\partial^2 + (B_x + D)\partial + 1/2\, D_x - A + s^2\varphi_x^2 B, \qquad (2.13)$$
$$L_- = -2s\varphi_x B\partial - (s\varphi_x D + 1/2\, D_x) - s((\varphi_x B)_x + i\varphi_t). \qquad (2.14)$$

To obtain the commutator, we need the following identities, which are valid for any operators $B \subset \mathcal{L}_m^1$, $N \in \mathcal{L}_p^2$:

$$[B\partial^2, N\partial] = [B, N]\partial^3 + (2BN_x - NB_x)\partial^2 + BN_{xx}\partial, \qquad (2.15)$$
$$[B\partial^2, N] = [B, N]\partial^2 + 2BN_x\partial + BN_{xx}, \qquad (2.16)$$
$$[B\partial, N\partial] = [B, N]\partial^2 + (BN_x - NB_x)\partial, \qquad (2.17)$$
$$[B\partial, N] = [B, N]\partial + BN_x.$$

All of them are proved by carying the operator ∂ through the operators B and N. From the form of the operators $L_\pm$ and formula (2.15) with $N = -2s\varphi_x B$, it follows that the coefficient of ∂^3 in the commutator $[L_+, L_-]$ is equal to zero. Therefore,

$$[L_+, L_-] = A_2\partial^2 + A_1\partial + A_0.$$

To derive the operator coefficients, we use formulas (2.13)–(2.17) and the equality

$$[i\partial_t, L_-] = -2si\varphi_{xt} B\partial - is\varphi_{xt} D - is((\varphi_x B)_x + i\varphi_t)_t,$$

which follows immediately from the form of the operator L_-. We have

$$A_2 = -4s\varphi_{xx} B^2 - 2s\varphi_x BB_x - s\varphi_x[B, D] - 1/2\,[B, D_x] - s\varphi_x[B, B_x], \qquad (2.18)$$

$$\begin{aligned}A_1 = &-2is\varphi_{xt} B - 2sB(\varphi_x B)_{xx} + 2B\{-s\varphi_x D - 1/2\, D_x - s((\varphi_x B)_x + i\varphi_t)\} \\ &+ (B_x + D)(-2s\varphi_x B)_x + 2s\varphi_x B(B_x + D)_x \\ &+ [B_x + D, -s\varphi_x D - 1/2\, D_x - s((\varphi_x B)_x + i\varphi_t)] \\ &+ [2s\varphi_x B, 1/2\, D_x - A + s^2\varphi_x^2 B]\end{aligned} \qquad (2.19)$$

$$\begin{aligned}A_0 = &-is\varphi_{xt} D - is((\varphi_x B)_x + i\varphi_t)_t \\ &+ B\{-(s\varphi_x D + 1/2\, D_x) - s((\varphi_x B)_x + i\varphi_t)\}_{xx} \\ &+ (B_x + D)\{-(s\varphi_x D + 1/2\, D_x) - s((\varphi_x B)_x + i\varphi_t)\}_x \\ &+ 2s\varphi_x B\{1/2\, D_x - A + s^2\varphi_x^2 B\}_x \\ &+ [1/2\, D_x - A + s^2\varphi_x^2 B, -(s\varphi_x D + 1/2\, D_x) - s((\varphi_x B)_x + i\varphi_t)].\end{aligned}$$

Chapter 2. A priori estimates and the uniqueness of solutions

From the last formula it follows that

$$A_0 = M + 2s^3 \varphi_x B(\varphi_x^2 B)_x + O(s\Lambda) + O(s^2), \qquad (2.20)$$

where the operator M is defined by equality (2.9). In the following, the operator of order $O(s\Lambda)$ denotes any operator $F : V \to V$ such that $|Fv| \leq cs|\Lambda v|$. Analogously, the term $O(s^2)$ denotes an operator F which admits the estimate $|Fv| \leq cs^2|v|$. The representation (2.19) implies $A_1 = O(s\Lambda)$. Thus, $A_1^* = O(s\Lambda)$ and therefore

$$|(A_1 \partial v, v)| = |(\partial v, A_1^* v)| \leq 2cs\|\partial v\| \, \|\Lambda v\| \leq c^2 \alpha s \|\partial v\|^2 + s\alpha^{-1} \|\Lambda v\|^2. \qquad (2.21)$$

Here we have used the α-inequality

$$2xy \leq \alpha x^2 + \alpha^{-1} y^2, \qquad x, y \geq 0, \quad al > 0.$$

We transform the formula $(A_2 \partial^2 v, v)$ by means of integrating by parts:

$$(A_2 \partial^2 v, v) = -(A_2 \partial v, \partial v) - (A_{2x} \partial v, v). \qquad (2.22)$$

In view of formula (2.18) and the fact that $[B, D], [B, D_x] \in \mathcal{L}_0^1$, we get

$$|A_{2x} v| \leq 2sc|v|$$

with some constant c, which implies

$$|(A_{2x} \partial v, v)| \leq 2cs\|\partial v\| \, \|v\| \leq c^2 s^2 \|v\|^2 + \|\partial v\|^2. \qquad (2.23)$$

Conditions (2.11) and (2.12) together with the above formulas yield

$$-(A_2, v_x, v_x) \geq 4s\mu(\varphi_{xx} v_x, v_x) - sc((|\varphi_x| + 1)v_x, v_x) \geq 2s\mu(\varphi_{xx} v_x, v_x)$$

if the number c_2 in estimate (2.12) is sufficiently large or, more precisely, $c_2 \geq c/2\mu$.

Combining this estimate with estimate (2.23), from formula (2.22) we get

$$\begin{aligned}(A_2 \partial^2 v, v) &\geq ((2s\mu\varphi_{xx} - 1)v_x, v_x) - c^2 s^2 \|v\|^2 \\ &\geq s\mu(\varphi_{xx} \partial v, \partial v) - c^2 s^2 \|v\|^2,\end{aligned} \qquad (2.24)$$

if $s \geq s_0$, where $4s_0 \mu = 2$, $c_2 \geq 1/\mu$. Finally, formula (2.20) and conditions (2.10)–(2.12) yield

$$\begin{aligned}(A_0, v, v) &\geq sc_0 \|\Lambda v\|^2 - sc_1 \|v\|^2 + 2s^3((2\varphi_x^2 \varphi_{xx} B^2 + \varphi_x^3 BB_x)v, v) \\ &\quad - \alpha c\|\Lambda v\|^2 - (\alpha^{-1} c + c)s^2 \|v\|^2.\end{aligned}$$

To estimate $O(s\Delta v, v)$, we have used the Cauchy–Schwarz–Bunyakovskii inequality and the α-inequality which yields

$$2s\|\Lambda v\|\,\|v\| \leq \alpha \|\Lambda v\|^2 + \alpha^{-1} s^2 \|v\|^2.$$

Set $\alpha = 1$. Then we obtain

$$2\varphi_x^2 \varphi_{xx} B^2 + \varphi_x^3 B B_x \geq 2\varphi_x^2(\mu\varphi_{xx} - c|\varphi_x|) \geq 2$$

if $c_2 \geq c/\mu$. Then, taking into account that $4s^3 - cs^2 \geq 2s^3$ and $sc_0 - c \geq sc_0/2$ for $s \geq s_0$, where s_0 is sufficiently large, we have

$$(A_0 v, v) \geq 2s^3 \|v\|^2 + s(c_0/2)\|\Lambda v\|^2. \tag{2.25}$$

Combining estimates (2.21), (2.24), and (2.25), we obtain

$$([L_+, L_-]v, v) \geq (2s^3 - c^2 s^2)\|v\|^2 + s((\mu\varphi_{xx} - c^2\alpha)\partial v, \partial v)$$
$$+ s(c_0/2 - \alpha^{-1})\|\Lambda v\|^2.$$

Suppose $\alpha^{-1} = c_0/4$ and let the number c_2 from condition (2.12) be chosen so that $c_2 \geq \alpha/\mu$. Then we have

$$([L_+, L_-]v, v) \geq s^3 \|v\|^2 + s\mu\|\partial v\|^2 + s(c_0/4)\|\Lambda v\|^2.$$

for sufficiently large s_0. Since $\|Pu\|_s^2 = ([L_+, L_-]v, v)$, returning to the function $u = e^{-s\varphi}v$ and acting as in the proof of Proposition 1.1, we obtain estimate (2.5) with some constant c. The theorem is proved. □

Condition (2.10) for the operator M for $D = 0$ and $B = E$ is transformed into condition (2.6) of Theorem 2.1. If $D = 0$, then Theorem 2.2 implies the following theorem.

Theorem 2.3. *Suppose that $A \in \mathcal{L}_2^1$, $B \in \mathcal{L}_0^3$, $[A, B] \in \mathcal{L}_1^0$ and for all $v \in V$ the following inequalities hold:*

$$\langle BA_x v, v\rangle \geq c_0|\Lambda v|^2 - c_1|v|^2, \qquad c_0 > 0,$$
$$\langle B^2 v, v\rangle \geq \mu|v|^2, \qquad \mu > 0.$$

If $-\varphi_x \geq 1$ and the inequality

$$\varphi_{xx} - c_2(|\varphi_x| + 1) \geq 1$$

holds for a sufficiently large number $c_2 = c_2(c_0, c_1, \mu)$, then there exist numbers $c, s_0 > 0$ such that for all $s \geq s_0$ and $u \in C_0^\infty(\Omega; V)$ estimate (2.5) holds.

Another particular case where condition (2.10) holds automatically is described in the following theorem.

Chapter 2. A priori estimates and the uniqueness of solutions

Theorem 2.4. *Under the assumptions of Theorem 2.2, let inequality (2.10) be replaced by the estimates*

$$\langle -D^2 v, v \rangle \geq c_0 |\Lambda v|^2 - c_1 |v|^2, \qquad c_0 > 0,$$

$$\langle ([A,D]-DD_x-2BA_x)v, v\rangle \leq m|\Lambda v|^2, \quad -\langle ([A,D_x]-DD_{xx})v,v\rangle \leq m|\Lambda v|^2,$$

and let $-\varphi_x \geq 1$. Then inequality (2.5) holds for all $u \in C_0^\infty(\Omega; V)$ and $s \geq s_0$, where s_0 is sufficiently large.

To prove this theorem, it suffices to note that

$$\langle Mv, v\rangle \geq s\varphi_{xx} c_0 |\Lambda v|^2 - s\varphi_{xx} c_1 |v|^2 - O(s|\varphi_x||\Lambda v|^2)$$
$$\geq sc_0 |\Lambda v|^2 - sc_3 |v|^2.$$

To obtain the last inequality, we have used condition (2.12), and took $c_3 = c_1 \max \varphi_{xx}$.

We denote by $C^2(\overline{\Omega}; V)$ the space of functions $u : \Omega \to H$ twice continuously differentiable in $\overline{\Omega}$ which take values in V together with their derivatives up to the second order. It is clear that, by continuity, estimate (2.5) is also valid for the functions of $C^2(\overline{\Omega}; V)$ that vanish on the boundary $\partial\Omega$ as well as their first derivatives. For simplicity, we assume that Ω has piecewise smooth boundary. As in Section 1, we restrict ourselves to the case where the domain Ω is the half of the ellipse

$$\Omega = \{x, t \in \mathbb{R} \mid \psi(x,t) = 1 - (x/r)^2 - (t/b)^2 > 0, \quad x > 0\}$$

and assume that the inequality

$$|Pu(x,t)| \leq c\left(|\Lambda u| + |\partial_x u|\right) + \left|\int_0^t (|\Lambda u(x,t)| + |\partial_x u(x,\tau)|)\,d\tau\right| \quad (2.26)$$

holds in Ω for a function $u \in C^2(\overline{\Omega}; V)$.

Theorem 2.5. *Let the assumptions of one of Theorems 2.1, 2.3 or 2.4 hold for an operator P. If a function $u \in C^2(\overline{\Omega}; V)$ satisfies inequality (2.26) and $u(0,t) = u_x(0,t) = 0$ for all $t \in (-b, b)$, then $u \equiv 0$ in Ω.*

Being similar to the proof of the uniqueness of the solution of inequality (1.39) used in Theorem 1.1, the proof of this theorem is left to the reader. We just note that the function

$$\varphi(x,t) = \exp(\lambda\psi(x+\delta,t) - 1)$$

introduced in Section 1 satisfies the assumptions of Theorems 2.1, 2.3, and 2.4 for sufficiently large $\lambda > 0$.

We consider now the examples of the application of Theorem 2.5 which will be used in the next chapter.

Example 2.1. Let a function $u \in C^\infty(\overline{K})$ satisfy the inequality

$$|iu_t + \Delta u| \leq c\left(|u| + |\nabla u| + \left|\int_0^t (|u(y,\tau)| + |\nabla u(t,\tau)|)\,d\tau\right|\right) \quad (2.27)$$

in the cylinder $K = \{y \in \mathbb{R}^2,\ t \in \mathbb{R} \mid |y| < 1\}$ and let

$$u = \frac{\partial u}{\partial |y|} = 0$$

on the boundary ∂K of the cylinder. Here Δ is the Laplace operator with respect to the variables y, and ∇ is the gradient with respect to y. We apply Theorem 2.5 to show that $u(x,t) = 0$ in K. To this end, we pass to the polar coordinates

$$y_1 = r\cos\varphi, \quad y_2 = r\sin\varphi, \quad r \in (0,1],\quad \varphi \in [0, 2\pi]$$

and set $x = 1 - r$. Then, for the variables x and φ, the Laplace operator takes the form

$$\Delta u = \partial^2 u + (1-x)^{-2}\partial_\varphi^2 u - (1-x)^{-1}\partial u$$

where $\partial = \partial/\partial x$. Elementary estimates show that for $x \in [0, 1-\varepsilon]$, $\varepsilon > 0$, inequality (2.27) implies

$$|iy_t + \partial^2 u + (1-x)^{-2}\partial_\varphi^2 u|$$
$$\leq c\left(|u| + |\partial u| + |\partial_\varphi u| + \left|\int_0^t (|u| + |\partial u| + |\partial_\varphi u|)\,d\tau\right|\right). \quad (2.28)$$

Let $H = L_2(0, 2\pi)$. We take the set of infinitely differentiable 2π-periodic functions as the linear manifold V. Define the operator $\Lambda : V \to V^2$ by the formula $\Lambda u = (\partial_\varphi u, u)$. For $u \in V$ we set

$$A(x)u = -(1-x)^{-2}\partial_\varphi^2 u, \qquad B(x) = E, \qquad D(x) = 0$$

where E is the unit operator. Then, upon integration with respect to φ from 0 to 2π, estimate (2.28) is transformed into estimate (2.26) with some constant c. Since

$$A'(x) = -2(1-x)^{-3}\partial_\varphi^2$$

Chapter 2. A priori estimates and the uniqueness of solutions   63

integrating by parts and taking into account the periodicity of $v \in V$, we obtain
$$\langle A'(x)v, v\rangle = 2(1-x)^{-3} \int_0^{2\pi} |\partial_\varphi v|^2 \, d\varphi \geq 2\varepsilon^{-3}|\Lambda v|^2.$$

Evidently, the operator A is symmetric and belongs to class $\mathcal{L}_2^1$ in the domain Ω from Theorem 2.5 with $r = 1 - \varepsilon$. Thus, the operator P satisfies the assumptions of Theorems 2.1, 2.5 and, consequently, $u \equiv 0$ in Ω. Since the numbers b and $1 - \varepsilon > 0$ which determine the semiaxes of the ellipse $\psi(x, t) = 0$ are arbitrary, the function u is equal to zero for all $x \in [0, 1]$, $t \in \mathbb{R}$, $\varphi \in [0, 2\pi]$, i.e., $u \equiv 0$ in $\overline{K}$.

Example 2.2. Let $K = \{x, y, t \in \mathbb{R} \mid x \in (0, r),\ y \in (0, 2\pi)\}$ and let a function $u \in C^\infty(\overline{K})$ be 2π-periodic in y and satisfy the inequality

$$|iu_t + \rho(x, y)(u_{xx} + u_{yy})|$$
$$\leq c\left(|u| + |u_x| + |u_y| + \left|\int_0^t (|u| + |u_x| + |u_y|)\, d\tau\right|\right). \quad (2.29)$$

Moreover, let $u(0, y, t) = u_x(0, y, t) = 0$. We show that if
$$\rho^2(x, y) \geq \mu, \qquad \partial_x(\rho^2(x, y)) \geq \mu,$$
where μ is a positive constant and $\rho \in C^\infty(\overline{K})$ is a real function which is 2π-periodic in y, then $u \equiv 0$ in K. We define the space H, the linear manifold V, and the operator Λ as in Example 2.1, the variable y standing for φ. Let $B(x)u = \rho(x, y)u$, $A(x)u = \partial_y(\rho(x, y)\partial_y u)$, and $D(x) \equiv 0$. Then inequality (2.29), possibly with another constant c, implies estimate (2.26), where
$$Pu = i\partial_t u + \partial(B(x)\partial u) - A(x)u, \qquad \partial = \partial/\partial x.$$

Evidently, $A \in \mathcal{L}_2^1$, $B \in \mathcal{L}_0^3$, and their commutator $[A, B] \in \mathcal{L}_1^0$. Further, if $\langle \cdot, \cdot \rangle$ is a scalar product in $L_2(0, 2\pi)$, then

$$\langle B^2 v, v\rangle = \int_0^{2\pi} \rho(x, y)|v(x, y)|^2 \, dy \geq \mu \int_0^{2\pi} |v(x, y)|^2 \, dy,$$

$$\langle BA_x v, v\rangle = -\int_0^{2\pi} \rho(x, y)\partial_y(\rho_x(x, y)\partial_y v)\overline{v}(x, y) \, dy$$
$$= \frac{1}{2}\int_0^{2\pi} (\rho^2)_x |v_y|^2 \, dy + \int_0^{2\pi} \rho_x \rho_y v_y \overline{v} \, dy$$
$$\geq c_0 \int_0^{2\pi} |v_y|^2 \, dy - c_1 \int_0^{2\pi} |v|^2 \, dy.$$

with some positive constants c_0 and c_1. Here we have integrated by parts to obtain the last inequality. To estimate the product $v_y \cdot \bar{v}$, we have used the α-inequality

$$2xy \leq \alpha x^2 + \alpha^{-1} y^2, \qquad \alpha, x, y > 0$$

with sufficiently small parameter α. These estimates show that the operator P satisfies the assumptions of Theorems 2.3, 2.5. Therefore, $u \equiv 0$ in Ω. Since the number B is arbitrary, $u \equiv 0$ in K.

Exercise 2.1. Generalize the result of Example 2.1 to the n-dimensional case.

Exercise 2.2. Prove the analogues of Theorems 2.1–2.5 for the operator P with the term $i\partial_t$ replaced by $-\partial_t^2$ or $-\partial_t$.

Exercise 2.3. Extend the results of Theorems 2.1–2.5 and Exercise 2.2 to the case where the operators A, B, and D depend on x and t.

Exercise 2.4. What is the minimal finite smoothness of the function u for which the results of Examples 2.1 and 2.2 still hold? To answer the question, analyse the proofs of Theorems 2.1–2.5.

2.3. REMARKS AND REFERENCES

The idea to prove the uniqueness of the solution of the Cauchy problem using an *a priori* estimate with a weight was set forth by Carleman (1939). Later this method was developed and generalized by many authors. General results are presented in Calderon (1958), Hörmander (1969), and Nirenberg (1975), where other related references can also be found. The method of weight *a priori* estimates was first used in the inverse problems of finding coefficients of differential equations in the Bukhgeim and Klibanov (1981). Propositions 1.1 and 1.2, and the examples from Section 2 are particular cases of the known estimates for the differential Schrödinger operator in $\mathbb{R}^n$ (see Isakov, 1980). The *a priori* estimates for partial differential equations with operator coefficients were obtained by the author and published for the first time. The fact that the problems of uniqueness for inverse problems can be reduced to the Cauchy problem for integro-differential equations was first noted in Beznoshchenko and Prilepko (1977).

Chapter 3.

Inverse problems for differential equations

3.1. ONE-DIMENSIONAL INVERSE PROBLEM FOR THE WAVE EQUATION IN LINEARIZED STATEMENT

One of the first approaches that arise when investigating a nonlinear inverse problem is to consider it in linearized statement. This approach is useful from two points of view. First, it often occurs that the sought object, for example, a coefficient of a differential equation, differs little from a certain known function. The problem linearized in the neighbourhood of this function may give satisfactory results from the viewpoint of applications. By contrast, application of the nonlinear algorithms provided by exact theory is too difficult or even gives worse results as far as stability and computation time are concerned. Second, the study of the linearized inverse problem often shows the way of solution in the initial nonlinear statement.

In this section, we consider the problem of reconstructing the index of refraction of a layered inhomogeneous medium in three-dimensional space. This problem provides a good illustration of the linearization method. For definiteness, we consider the acoustic interpretation of the wave equation. For the wave process taking place in half-space, the corresponding boundary value problem is considered with the Dirichlet or Neumann conditions. The point source of sound is assumed to be concentrated in the neighbourhood of the boundary. These two variants of the inverse problem are formulated in the exercises as well as some multidimensional inverse problems in linearized statement.

We begin with the general statement of the problem

Let a function $u(x,t)$, $x \in \mathbb{R}^3$, $t \in \mathbb{R}$, be a solution of the generalized Cauchy problem

$$n^2(x_3)u_{tt} - c_0^2 \Delta u = f(x,t), \tag{1.1}$$

$$u|_{t<0} \equiv 0. \tag{1.2}$$

Here $f(x,t)$ is a generalized function that defines the distribution of acoustic sound sources in the space and the law of their functioning in time. Since the medium is assumed to be quiescent for $t < 0$,

$$f(x,t)|_{t<0} \equiv 0.$$

The function $u(x,t)$ determines the sound pressure at the moment t at the point $x = (x_1, x_2, x_3) \in \mathbb{R}^3$. The function $n(x_3) = c_0/c(x_3)$ is the index of refraction of the medium; $c(x_3)$ is the sound velocity, which depends only on x_3; $c_0 = c(x_3^0)$ is the sound velocity at a fixed point $x^0 = (x_1^0, x_2^0, x_3^0)$. The symbol Δ denotes the Laplace operator

$$\Delta = \sum_{j=1}^{3} \partial^2/\partial x_j^2.$$

Given the function $\varphi(t) = u(x^1, t)$, where $x^1 = (x_1^1, x_2^1, x_3^1)$ is a fixed point in $\mathbb{R}^3$, it is required to reconstruct the index of refraction $n(x_3)$ or, what is the same, the speed $c(x_3)$ in the corresponding interval. In many kinds of real media (for example, in the water or in the air) the sound velocity varies over a small range (for example, in the water $c = 1500 \pm 150$ m/c). Therefore, it is natural to seek the index of refraction in the form

$$n(x_3) = 1 + a(x_3) \tag{1.3}$$

where $|a(x_3)| \ll 1$, and neglect the terms of order $o(a)$ in calculations. Thus we arrive at the linearized statement of the inverse problem. Following expansion (1.3), we shall seek the solution of problem (1.1), (1.2) in the form

$$u = u_0 + v, \tag{1.4}$$

where the function u_0 describes the wave process in a nonperturbed medium corresponding to $a \equiv 0$:

$$u_{0tt} - c_0^2 \Delta u_0 = f(x,t), \tag{1.5}$$

Chapter 3. Inverse problems for differential equations

$$u_0|_{t<0} \equiv 0. \tag{1.6}$$

From (1.1) and (1.4) we derive the equation for the function v:

$$(1 + 2a + a^2)(u_{0tt} + v_{tt}) - c_0^2 \Delta u_0 - c_0^2 \Delta v = f(x,t),$$

from which, omitting the terms of order $o(a)$ and taking into account (1.5) and the initial conditions (1.2), (1.6), we obtain

$$v_{tt} - c_0^2 \Delta v = -2a(x_3) u_{0tt},$$

$$v|_{t<0} \equiv 0.$$

The function v represents the wave scattered on the inhomogeneity of the index of refraction. Using the representation (1.4), we make the statement of the inverse problem more precise. It is required to reconstruct $a(x_3)$ by the scattered wave

$$v(x^1, t) = \psi(t) \tag{1.7}$$

measured in a point x^1, where $\psi(t) = \varphi(t) - u_0(x^1, t)$.

Suppose $w(x,t)$ is a fundamental solution of the wave equation with speed c_0:

$$\Box_{c_0} w \equiv w_{tt} - c_0^2 \Delta w = \delta(x,t), \qquad x \in \mathbb{R}^3, \quad t \in \mathbb{R}.$$

Here $\delta(x,t)$ is the Dirac function concentrated in the point $x = 0$, $t = 0$.
It is known (Vladimirov, 1971, p. 200) that

$$w(x,t) = \frac{\Theta(t)}{2\pi c_0} \delta(c_0^2 t^2 - |x|^2), \tag{1.8}$$

where $\Theta(t)$ is the Heaviside function and $\delta(p)$ is the Dirac function concentrated on the surface $p = c_0^2 t^2 - |x|^2 = 0$, which, in our case, is a circular cone. Also, it is known that the solution of the generalized Cauchy problem

$$\Box_{c_0} v = F(x,t),$$

$$v|_{t<0} \equiv 0,$$

where $F \in \mathcal{D}'(\mathbb{R}^4)$, $F = 0$ for $t < 0$, is the convolution of the function F with the fundamental solution w (Vladimirov, 1971, p. 225):

$$v(x,t) = (w * F)(x,t) = \int F(\xi, \tau) w(x - \xi, t - \tau) \, d\xi \, d\tau.$$

In particular, setting $x = x^1$, $F(x,t) = -2a(x_3)u_{0tt}$, we obtain

$$\psi(t) = v(x^1, t) = -2 \int a(\xi_3) u_{0tt}(\xi, \tau) w(x^1 - \xi, t - \tau) \, d\xi \, d\tau, \qquad (1.9)$$

or, in the short form,

$$\psi(t) = v(x^1, t) = w * F|_{x=x^1}.$$

For the study of this integral equation it is convenient to use the following representation of the fundamental solution $w(x,t)$:

$$w(x,t) = (4\pi c_0)^{-1} \delta(c_0 t - |x|) |x|^{-1}. \qquad (1.10)$$

To obtain it from (1.8), we use the following property:

$$\delta(\lambda p(x)) = |\lambda|^{-1} \delta(p(x)), \qquad (1.11)$$

where $\lambda \neq 0$, $\nabla p(x) = \operatorname{grad} p(x) \neq 0$ for $p(x) = 0$. The property (1.11) implies that the Dirac function is a homogeneous generalized function with homogeneity degree $\alpha = -1$. Using this property for $t > 0$, we obtain

$$\delta(c_0^2 t^2 - |x|^2) = \delta((c_0 t + |x|)(c_0 t - |x|))$$
$$= \delta(c_0 t - |x|)(c_0 t + |x|)^{-1} = \delta(c_0 t - |x|)(2|x|)^{-1} \qquad (1.12)$$

because $c_0 t + |x| = 2|x|$ on the surface $p \equiv c_0 t - |x| = 0$. Observing that $\delta(c_0 t - |x|) = 0$ for $t < 0$, we may omit the multiplier $\Theta(t)$. Therefore, (1.8) and (1.12) imply (1.10).

Another deduction of equality (1.10), which is less formal, consists in finding the convolution $F * w$ and comparing the result obtained with the classical Kirchhoff formula. From (1.10) we obtain

$$(F * w)(x, t) = \int \frac{F(\xi, \tau)}{4\pi c_0} \frac{\delta(c_0(t - \tau) - |x - \xi|)}{|x - \xi|} \, d\xi \, d\tau. \qquad (1.13)$$

The following general formula serves for calculating the integrals which contain the delta function concentrated on a smooth nondegenerate surface of dimension $n - k$ in $\mathbb{R}^n$:

$$\int_{\mathbb{R}^n} u(x) \delta(p(x)) \, dx = \int_{p=0} u(x) |\partial p(x)|^{-1} \, d\sigma. \qquad (1.14)$$

Here $u \in C_0^\infty(\mathbb{R}^n)$, $p(x) = (p_1(x), p_2(x), \ldots, p_k(x))$, $k \geq n$, $p_j \in C^\infty(\mathbb{R}^n)$, $j = 1, \ldots, k$. We assume that the vectors $\partial p_j(x) = (\partial_1 p_j, \ldots, \partial_n p_j)$, where $\partial_j = \partial/\partial x_j$, are linearly independent on the surface $p(x) = 0$. By this assumption, the determinant of the Gram matrix $A = (a_{ij})$ composed of the scalar products $a_{ij} = \langle \partial p_i, \partial p_j \rangle$ is positive. The quantity

$$|\partial p(x)| = (\det A)^{1/2}$$

is the volume of the parallelepiped formed by the vectors ∂p_j, and $d\sigma$ is the Euclidean element of the area of the surface $p = 0$. Setting

$$p(\tau) = c_0 t - c_0 \tau - |x - \xi|$$

and observing that $\partial_\tau p = -c_0$, $|\partial_\tau p| = c_0$, using formula (1.14) we may avoid integration with respect to the variable τ in formula (1.13). As a result we obtain the Kirchhoff formula

$$(F * w)(x, t) = \frac{1}{4\pi c_0^2} \int \frac{F(\xi, t - |x - \xi| c_0^{-1})}{|x - \xi|} d\xi$$

$$= \frac{1}{4\pi c_0^2} \int_{|x-\xi|\leq c_0 t} \frac{F(\xi, t - |x - \xi| c_0^{-1})}{|x - \xi|} d\xi.$$

In the last equality, we have used the fact that $F(x, t) = 0$ for $t < 0$. Thus, formula (1.10) is established.

Now, we transform formula (1.9). Since

$$F(x, t) = -\partial_t^2 (2a(x_3) u_0(x, t))$$

and $v = F * w$ in (1.9), using the rule of differentiating the convolution

$$\partial^\alpha (u * v) = (\partial^\alpha u) * v = u * (\partial^\alpha v),$$

we get

$$v = \partial_t^2 (F_0 * w), \qquad F_0 = -2a(x_3) u_0(x, t)$$

or, in detail,

$$v(x, t) = -2\partial_t^2 \int a(\xi_3) u_0(\xi, \tau) W(x - \xi, t - \tau) \, d\xi \, d\tau. \qquad (1.15)$$

For further transformations it is necessary to specify the form of the right-hand side $f(x, t)$ in equation (1.1) or, in other words, the form of the source.

First, we suppose that

$$f(x, t) = m\delta(x - x^0)\delta(t).$$

Then $u_0 = f * w$,

$$u_0(\xi, \tau) = \int f(y,s) w(\xi - y, \tau - s)\, dy\, ds$$
$$= m \int \delta(y - x^0) \delta(s) w(\xi - y, \tau - s)\, dy\, ds$$
$$= mw(\xi - x^0, \tau) = m(4\pi c_0)^{-1} \delta(c_0 \tau - |\xi - x^0|)|\xi - x^0|^{-1}.$$

Substituting this formula into (1.15), we obtain

$$v(x,t) = -\frac{2m}{(4\pi c_0)^2} \partial_t^2 \int a(\xi_3) \frac{\delta(c_0 \tau - |\xi - x^0|)}{|\xi - x^0|} \frac{\delta(c_0(t-\tau) - |x - \xi|)}{|x - \xi|}\, d\xi\, d\tau$$
$$= -\frac{2m}{(4\pi c_0)^2 c_0} \partial_t^2 \int a(\xi_3) \frac{\delta(c_0 t - |\xi - x^0| - |x - \xi|)}{|\xi - x^0||\xi - x|}\, d\xi. \quad (1.16)$$

In the last equality we avoided the integration with respect to τ using formula (1.14).

We set

$$K(t, \xi_3) = K(t, \xi_3, x^0, x^1) = \int \frac{\delta(c_0 t - |\xi - x^0| - |\xi - x^1|)}{|\xi - x^0||\xi - x^1|}\, d\xi_1\, d\xi_2. \quad (1.17)$$

Then, (1.16), (1.17), and (1.7) yield

$$\psi(t) = v(x^1, t) = -\frac{2m}{(4\pi c_0)^2 c_0} D_t^2 \int K(t, \xi_3) a(\xi_3)\, d\xi_3. \quad (1.18)$$

Thus, we have obtained the one-dimensional integro-differential equation for determining small variations of the index of refraction $a(x_3)$ by the scattered wave $\psi(t)$ known at the point x^1. The properties of this kernel $K(t, \xi_3, x^0, x^1)$ essentially depend on the relative position of the points x^0 and x^1, that is, of the receiver and the source. First, we consider the case

$$x_j^0 = x_j^1, \quad j = 1, 2.$$

As the problem in question is invariant with respect to the shift in x_1 and x_2, without loss of generality we may assume that

$$x_j^0 = x_j^1 = 0, \quad j = 1, 2.$$

Chapter 3. Inverse problems for differential equations

In this case, the surface

$$p(\xi) = c_0 t - |\xi - x^0| - |\xi - x^1| = 0$$

is an ellipsoid of revolution with focuses $(0, 0, x_3^0)$, $(0, 0, x_3')$ on the ξ_3-axis. The integral (1.17) is taken along the circle formed by the intersection of the said ellipsoid with the plane $\xi_3 = $ const (see Figure 5). From the geometrical considerations it is evident that the integral K is not equal to zero only when $\xi_3 \in (A, B)$, i.e., for $\xi_3 \notin (A, B)$

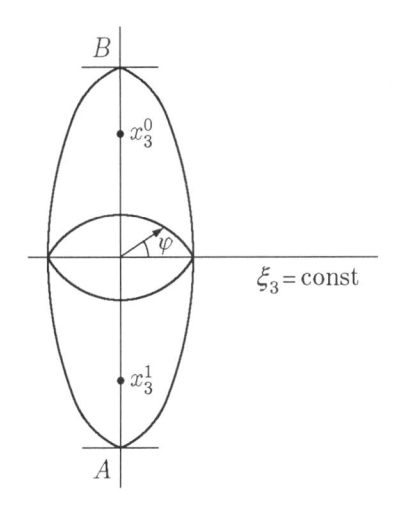

$$K(t, \xi_3) = 0. \qquad (1.19)$$

Figure 5

To calculate $K(t, \xi_3)$ for $\xi_3 \in (A, B)$ we pass to the polar coordinate system

$$\xi_1 = r \cos\varphi, \qquad \xi_2 = r \sin\varphi, \qquad d\xi_1\, d\xi_2 = r\, dr\, d\varphi,$$

$$|\xi - x^0| = \sqrt{r^2 + (\xi_3 - x_3^0)^2}, \qquad |\xi - x^1| = \sqrt{r^2 + (\xi_3 - x_3^1)^2},$$

$$p(\xi) = c_0 t - \sqrt{r^2 + (\xi_3 - x_3^0)^2} - \sqrt{r^2 + (\xi_3 - x_3^1)^2} = p(r).$$

From the geometrical point of view it is evident that the equation $p(r) = 0$ for $\xi_3 \in (A, B)$ has unique positive simple root $r = r^*$. Therefore, taking into account the one-dimensional variant of formula (1.14)

$$\int_0^\infty u(r)\delta(p(r))\, dr = \frac{u(r^*)}{|p'(r^*)|},$$

which is valid in the case of the unique positive simple root $r = r^*$ of the equation $p(r) = 0$, $r \in \mathbb{R}$, we have

$$K(t, \xi_3) = \int_0^{2\pi} d\varphi \int_0^\infty \frac{r\delta(p(r))\, dr}{\sqrt{r^2 + (\xi_3 - x_3^0)^2}\sqrt{r^2 + (\xi_3 - x_3^1)^2}} = \frac{2\pi u(r^*)}{p'(r^*)},$$

where

$$u(r) = \frac{r}{\sqrt{r^2 + (\xi_3 - x_3^0)^2}\sqrt{r^2 + (\xi_3 - x_3^1)^2}}.$$

Since
$$p'_r(r) = -\frac{r}{\sqrt{r^2 + (\xi_3 - x_3^0)^2}} - \frac{r}{\sqrt{r^2 + (\xi_3 - x_3^1)^2}}$$
$$= -\frac{r\left[\sqrt{r^2 + (\xi_3 - x_3^0)^2} + \sqrt{r^2 + (\xi_3 - x_3^1)^2}\right]}{\sqrt{r^2 + (\xi_3 - x_3^0)^2}\sqrt{r^2 + (\xi_3 - x_3^1)^2}},$$

taking into account the equality
$$c_0 t = \sqrt{r^2 + (\xi_3 - x_3^0)^2} + \sqrt{r^2 + (\xi_3 - x_3^1)^2},$$
for $p(r) = 0$, i.e., $r = r^*$, we find
$$p'_r(r^*) = -\frac{rc_0 t}{\sqrt{r^2 + (\xi_3 - x_3^0)^2}\sqrt{r^2 + (\xi_3 - x_3^1)^2}}\bigg|_{r=r^*}.$$

Therefore,
$$\frac{u(r^*)}{|p'_r(r^*)|} = (c_0 t)^{-1}$$
and
$$K(t, \xi_3) = 2\pi(c_0 t)^{-1}, \qquad \xi_3 \in (A, B). \tag{1.20}$$

From Figure 5 it is evident that
$$B = \frac{x_3^0 + x_3^1 + c_0 t}{2}, \qquad A = \frac{x_3^0 + x_3^1 - c_0 t}{2}.$$

Therefore, (1.19) and (1.20) imply
$$\int K(t, \xi_3) a(\xi_3)\, d\xi_3 = \frac{2\pi}{c_0 t} \int_{(x_3^0 + x_3^1 - c_0 t)/2}^{(x_3^0 + x_3^1 + c_0 t)/2} a(\xi_3)\, d\xi_3.$$

As a result, equation (1.18) assumes the form
$$-\frac{4\pi}{(4\pi c_0)^2 c_0^2} \partial_t^2 \frac{1}{t} \int_{(x_3^0 + x_3^1 - c_0 t)/2}^{(x_3^0 + x_3^1 + c_0 t)/2} a(\xi_3)\, d\xi_3 = \psi(t). \tag{1.21}$$

Here we have assumed that
$$c_0 t > |x_3^0 - x_3^1|,$$
which means that the ellipsoid in Figure 5 is nondegenerate. To simplify formula (1.21), we assume that
$$x^0 = x^1 = 0,$$

Chapter 3. Inverse problems for differential equations

and for $\xi_3 \leq 0$
$$a(\xi_3) = 0, \tag{1.22}$$
$$a'(0) = 0. \tag{1.23}$$

Then
$$D_t^2 \frac{1}{t} \int_0^{c_0 t/2} a(\xi_3)\, d\xi_3 = \rho\psi(t), \tag{1.24}$$
$$\rho = -\frac{(4\pi c_0)^2 c_0^2}{4\pi m}.$$

Now, we put
$$(I\psi)(t) = \int_0^t \psi(\tau)\, d\tau.$$

Applying the integration operator I twice to equality (1.24) and taking (1.22) and (1.23) into account, we obtain
$$\frac{1}{t}\int_0^{c_0 t/2} a(\xi_3)\, d\xi_3 = \rho I^2 \psi.$$

Hence, multiplying by t and differentiating, we find
$$a(c_0 t/2) \cdot c_0/2 = \rho \partial_t (tI^2 \psi).$$

Since
$$(I^2\psi)(t) = \int_0^t (t-\tau)\psi(\tau)\, d\tau,$$

we have
$$\partial_t(tI^2\psi) = \partial_t \int_0^t t(t-\tau)\psi(\tau)\, d\tau = \int_0^t (2t-\tau)\psi(\tau)\, d\tau,$$

and, consequently,
$$a(c_0 t/2) = \frac{2\rho}{c_0} \int_0^t (2t-\tau)\psi(\tau)\, d\tau.$$

When $m = 4\pi c_0^2$, the constant $2c_0^{-1}\rho$ takes the form $-2c_0$. Therefore, for such value of the power m we have
$$a(c_0 t/2) = -2c_0 \int_0^t (2t-\tau)\psi(\tau)\, d\tau. \tag{1.25}$$

Thus, we have obtained an explicit formula of inversion which connects the variation of the index of refraction $a(x_3)$ with the scattered wave $\psi(t)$ measured in a fixed point.

74   A. L. Bukhgeim. Introduction to the theory of inverse problems

Exercise 1.1. Suppose that the wave process described by conditions (1.1) and (1.2) takes place in the half-space $x_3 > 0$ and the Neumann condition $\partial u/\partial x_3 = 0$ holds on its boundary. Suppose the right-hand side is as follows:
$$f(x,t) = 4\pi c_0^2 \delta(x - x^0)\delta(t), \qquad x^0 = (0,0,h), \quad h > 0.$$

Setting $x^1 = x^0$, $\psi(t) = v(x^1, t)$ and passing to the limit as $h \to 0$, obtain the inversion formula
$$a(c_0 t/2) = -8c_0 \int_0^t (2t - \tau)\psi(\tau)\,d\tau$$
in the linear approximation under the condition that $a(0) \equiv a'(0) = 0$.

Exercise 1.2. In the Cauchy problem (1.1), (1.2), suppose that
$$f(x,t) = 4\pi c_0^2 [\delta(x - x^0) - \delta(x + x^0)]\delta(t), \qquad x^0 = (0,0,h), \quad h > 0,$$
$n(x_3) = 1 + a(x_3)$ is an even function, $a^{(k)}(0) = 0$, $k = 0,1,2,3$. Let $\psi(t) = v'_{x_3}(0,t)$, where the function v is defined by formula (1.4). Passing to the limit as $h \to 0$, prove the inversion formula
$$a(c_0 t/2) = -2c_0^3 \int_0^t \tau \int_0^\tau (\tau - s)\psi(s)\,ds\,d\tau$$
in the linear approximation. Show that this formula provides the solution of inverse problem (1.1), (1.2) in linearized statement in the half-space $x_3 > 0$ with Dirichlet condition $u = 0$ on the boundary $x_3 = 0$ and with right-hand side $f = 4\pi c_0^2 \delta'_{x_3}(x)\delta(t)$.

Exercise 1.3. Let $\mathbb{R}^3_+ = \mathbb{R}^3 \cap \{x \mid x_3 > 0\}$, $x' = (x_1, x_2)$, $a \in C_0^\infty(\mathbb{R}^3)$,
$$(A_c a)(x', t) = -2\pi ct \int_{\mathbb{R}^3_+} \delta'(|x' - y'|^2 + y_3^2 - c^2 t^2)a(y)\,dy = b(x', t).$$

Show that the unique solution of the equation $A_c a = b$ is given by the formula
$$a(x) = (A_c^{-1} b)(x) = -2\pi c x_3 \int_{\mathbb{R}^3_+} \delta'(|x' - y'|^2 + x_3^2 - c^2 t^2)b(y', t)\,dy'\,dt.$$

(Directions: see Garipov and Kardakov (1973), Bukhgeim and Kardakov (1978)).

Chapter 3. Inverse problems for differential equations

Exercise 1.4. Consider the following boundary value problem

$$n^2(x)u_{tt} - \Delta u = \delta(x - x^0)\delta(t), \qquad x, x^0 \in \mathbb{R}^3_+, \qquad (1.26)$$

$$u|_{t<0} \equiv 0, \qquad u_{x_3}|_{x_3=0} = 0. \qquad (1.27)$$

Here $n(x) = 1 + a(x)$, $a \in C_0^\infty(\mathbb{R}^3_+)$, $|a| \ll 1$,

$$u = u_0(x, x^0, t) + v(x, x^0, t),$$

where u_0 is a solution of the problem (1.26), (1.27) for

$$a(x) \equiv 0, \qquad \psi(x', t) = v(x, x, t)|_{x_3=0}.$$

The function ψ is a wave scattered in the inverse direction. Show that

$$a(x) = 4\pi^2 A_{1/2}^{-1} \partial_t(tI^2\psi)$$

in the linear approximation, where

$$(I\psi)(x', t) = \int_0^t \psi(x', \tau)\, d\tau$$

and the operator $A_{1/2}$ is defined in Exercise 1.3.

3.2. THE METHOD OF TRANSFORMATION OPERATORS

In this section we establish the similarity of the Sturm–Liouville operators on the semiaxis $x \geq 0$ with a given boundary condition for $x = 0$. This result is used later in the proof of the continuous analogue of Theorem 3.7, Chapter 1. We shall use this result also in Section 3, Chapter 4, when investigating the stability of the nonhyperbolic Cauchy problem for the wave equation.

So, we suppose that

$$A_j = -\partial^2 + a_j(x), \qquad \partial = \partial/\partial x$$

are two differential operators given for $x \geq 0$ with continuous complex coefficients $a_j(x)$ and with domains of definition

$$D(A_j) = \{u \in C^2[0, \infty) \mid u'(0) - h_j u(0) = 0\}.$$

Here h_j are complex numbers in the general case. We aim to construct an operator F such that

$$A_2 F u = F A_1 u, \qquad \forall u \in \mathcal{D}(A_1), \qquad (2.1)$$

$$F : \mathcal{D}(A_1) \to \mathcal{D}(A_2). \qquad (2.2)$$

We shall seek the operator F satisfying conditions (2.1) and (2.2) in the form

$$(Fu)(x) = u(x) + \int_0^x K(x,t) u(t)\, dt, \qquad (2.3)$$

where $K(x,t)$ is the unknown smooth kernel of the Volterra operator K. Since the operator $E + K$ has bounded inverse operator of the same form, we have

$$F^{-1} u = u(x) + \int_0^x L(x,t) u(t)\, dt.$$

Therefore, equality (2.1) may be rewritten as follows:

$$A_1 = F^{-1} A_2 F.$$

Thus, the operators A_j, $j = 1, 2$ are similar. The operator F satisfying conditions (2.1) and (2.2) is said to *interlace the operators A_1 and A_2*. It is also called *the transformation operator*.

We shall now seek the function $K(x,t)$. Since

$$\partial F u(x) = \partial u(x) + K(x,x) u(x) + \int_0^x \partial_x K(x,t) u(t)\, dt \qquad (2.4)$$

and for $u \in \mathcal{D}(A_1)$ it is required that $Fu \in \mathcal{D}(A_2)$, setting $x = 0$ in (2.3) and (2.4), we have

$$(\partial F u - h_2 F u)(0) = \partial u(0) + (K(0,0) - h_2) u(0) = (h_1 - h_2 + K(0,0)) u(0),$$

if $u \in \mathcal{D}(A_1)$. Thus, condition (2.2) holds if

$$K(0,0) = h_2 - h_1. \qquad (2.5)$$

Differentiating formula (2.4) one more time, we have

$$(Fu)'' = u''(x) + K(x,x) u'(x) + u(x) \frac{dK(x,x)}{dx} + \partial_x K|_{t=x} u(x)$$
$$+ \int_0^x \partial_x^2 K(x,t) u(t)\, dt,$$

Chapter 3. Inverse problems for differential equations

which implies

$$A_2 F u = -u'' - K(x,x)u'(x) - u\frac{\mathrm{d}K(x,x)}{\mathrm{d}x} - \partial_x K|_{t=x} u(x)$$
$$+ a_2(x)u + \int_0^x (-K_{xx} + a_2(x)K)u(t)\,\mathrm{d}t. \tag{2.6}$$

On the other hand, transforming the expression $FA_1 u$ by means of integrating by parts, we obtain

$$FA_1 u = -u'' + a_1(x)u + \int_0^x K(x,t)(-u_{tt}(t) + a_1(t)u(t))\,\mathrm{d}t$$
$$= -u'' + a_1 u - K u_t|_0^x + \int_0^x [K_t(x,t)u_t(t) + a_1(t)K(x,t)u(t)]\,\mathrm{d}t$$
$$= -u'' + a_1 u - K(x,x)u'(x) + K(x,x)u'(x)$$
$$+ K_t(x,x)u(x) - K_t(x,0)u(0) + \int_0^x (-K_{tt} + a_1(t)K)u(t)\,\mathrm{d}t.$$

Equating this expression to (2.6) and observing that the function $u \in D(A_1)$ is arbitrary, we get

$$K_{xx} - a_2(x)K = K_{tt} - a_1(t)K, \qquad x \geq t \geq 0, \tag{2.7}$$

$$2\frac{\mathrm{d}K(x,x)}{\mathrm{d}x} = a_2(x) - a_1(x), \qquad x \geq 0, \tag{2.8}$$

$$(K_t - h_1 K)|_{t=0} = 0, \qquad x \geq 0. \tag{2.9}$$

To obtain formula (2.8), we have used the fact that

$$\frac{\mathrm{d}K(x,x)}{\mathrm{d}x} = (K_x + K_t)|_{t=x}.$$

The boundary condition $u'(0) = h_1 u(0)$ was used to deduce formula (2.9). Integrating condition (2.8) with respect to x from 0 to x and taking into account equality (2.5), we have

$$K(x,x) = h_2 - h_1 + \frac{1}{2}\int_0^x (a_2(s) - a_1(s))\,\mathrm{d}s. \tag{2.10}$$

Evidently, condition (2.10) is equivalent to conditions (2.5) and (2.8).

To prove the existence of solutions of the problem (2.7), (2.9), (2.10) we assume that $a_j \in C^1[0,\infty)$ and make the change of variables

$$\xi = x + t, \qquad \eta = x - t, \qquad K(x,y) = v(\xi,\eta).$$

Then
$$K_x = v_\xi + v_\eta, \qquad K_{xx} = v_{\xi\xi} + 2v_{\xi\eta} + v_{\eta\eta}$$
$$K_t = v_\xi - v_\eta, \qquad K_{tt} = v_{\xi\xi} - 2v_{\xi\eta} + v_{\eta\eta}$$

and, consequently, equation (2.7) takes the form

$$v_{\xi\eta} = \frac{1}{4} q(\xi, \eta) v(\xi, \eta), \qquad (2.11)$$

where $q(\xi, \eta) = a_2(x) - a_1(t)$. The boundary conditions are transformed as follows:

$$K(x, x) = v(2x, 0) = v(\xi, 0) = h_2 - h_1 + \frac{1}{2} \int_0^{\xi/2} q(s, 0)\, ds \qquad (2.12)$$

$$K_t(x, 0) - h_1 K(x, 0) = v_\xi(\xi, \xi) - v_\eta(\xi, \xi) - h_1 v(\xi, \xi) = 0. \qquad (2.13)$$

Integrating equation (2.11) with respect to η from zero to η and taking into account that (2.12) implies $v_{\xi,0} = 1/4\, q(\xi/2, 0)$, we have

$$v_\xi(\xi, \eta) = \frac{1}{4} \int_0^\eta q(\xi, \alpha) v(\xi, \alpha)\, d\alpha + \frac{1}{4} q\left(\frac{\xi}{2}, 0\right). \qquad (2.14)$$

Now, integrating with respect to ξ from η to ξ, we obtain

$$v(\xi, \eta) = \frac{1}{4} \int_\eta^\xi d\beta \int_0^\eta q(\beta, \alpha) v(\beta, \alpha)\, d\alpha + \frac{1}{4} \int_\eta^\xi q\left(\frac{\beta}{2}, 0\right) d\beta + v(\eta, \eta). \qquad (2.15)$$

To find $v(\eta, \eta)$ we note that the boundary condition (2.13) may be rewritten as follows:

$$2v_\xi(\eta, \eta) e^{h_1 \eta} = (h_1 v(\eta, \eta) + v_\xi(\eta, \eta) + v_\eta(\eta, \eta)) e^{h_1 \eta} = (e^{h_1 \eta} v(\eta, \eta))'_\eta. \qquad (2.16)$$

On the other hand, following (2.14), we obtain

$$2v_\varepsilon(\eta, \eta) e^{h_1 \eta} = \frac{1}{2} \int_0^\eta e_1^h \eta q(\eta, \alpha) v(\eta, \alpha)\, d\alpha + \frac{1}{2} q\left(\frac{\eta}{2}, 0\right) e^{h_1 \eta}. \qquad (2.17)$$

Integrating formula (2.17) with respect to η from 0 to η and taking into account (2.16), we obtain

$$e^{h_1 \eta} v(\eta, \eta) - v(0, 0) = \frac{1}{2} \int_0^\eta d\beta \int_0^\beta e^{h_1 \beta} q(\beta, \alpha) v(\beta, \alpha)\, d\alpha$$
$$+ \frac{1}{2} \int_0^\eta e^{h_1 \beta} q\left(\frac{\beta}{2}, 0\right) d\beta.$$

Chapter 3. Inverse problems for differential equations

Since $v(0,0) = h_2 - h_1$ (see (2.12)), we finally obtain

$$v(\eta,\eta) = e^{h_1\eta}(h_2 - h_1) + \frac{1}{2}\int_0^\eta d\beta \int_0^\beta e^{-h_1(\eta-\beta)} q(\beta,\alpha) v(\beta,\alpha)\, d\alpha$$
$$+ \frac{1}{2}\int_0^\eta e^{-h_1(\eta-\beta)} q\left(\frac{\beta}{2}, 0\right) d\beta. \qquad (2.18)$$

From formulas (2.15) and (2.18) it follows that v is a solution of the integral equation

$$v = Qv + f, \qquad (2.19)$$

where the right-hand side f is defined by the formula

$$f(\xi,\eta) = \frac{1}{4}\int_\eta^\xi q\left(\frac{\beta}{2}\right) d\beta + e^{-h_1\eta}(h_2 - h_1) + \frac{1}{2}\int_0^\eta e^{-h_1(\eta-\beta)} q\left(\frac{\beta}{2}, 0\right) d\beta, \qquad (2.20)$$

and the operator Q acts as follows:

$$(Qv)(\xi,\eta) = \frac{1}{4}\int_\eta^\xi d\beta \int_0^\eta q(\beta,\alpha) v(\beta,\alpha)\, d\alpha$$
$$+ \frac{1}{2}\int_0^\eta d\beta \int_0^\beta e^{-h_1(\eta-\beta)} q(\beta,\alpha) v(\beta,\alpha)\, d\alpha. \qquad (2.21)$$

If we want to construct the function $K(x,t)$ in the domain $T > x > t > 0$, then we need to find the function $v(\xi,\eta)$ in the domain

$$\Omega = \{\xi,\eta \mid 2T > \xi + \eta,\ \ \xi > \eta > 0\}$$

shown in Figure 6.

We shall seek a solution of equation (2.19) in the space $C(\overline{\Omega})$, where a set of equivalent norms is introduced as follows:

$$\|v\|_s = \sup_\Omega e^{-s\eta} |v(\xi,\eta)|, \qquad s \geq 0.$$

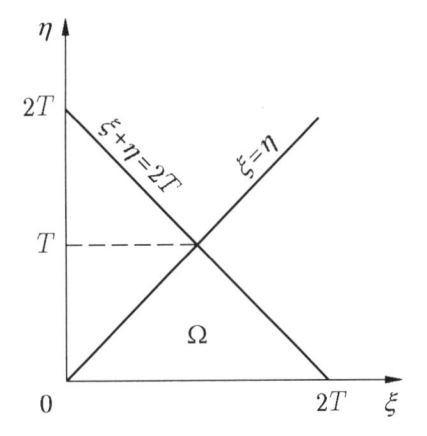

Figure 6

We show that for sufficiently large s_0 we have

$$\|Qv\|_s \leq 1/2\, \|v\|_s, \qquad s \geq s_0, \quad v \in C(\overline{\Omega}). \qquad (2.22)$$

Therefore, equation (2.19) has a unique solution $v \in C(\overline{\Omega})$, which may be obtained using the method of successive approximations. From the definition of $\|v\|_s$ and the operator Q, for $|q(\xi, \eta)| \leq c$ we have

$$e^{-s\eta}|Qv(\xi, \eta)|$$
$$\leq \frac{1}{4}c \int_\eta^\xi d\beta \int_0^\eta e^{-s(\eta-\alpha)} e^{-s\alpha}|v(\beta, \alpha)|\, d\alpha$$
$$+ \frac{1}{2}c \int_0^e ta_0\, d\beta \int_0^\beta e^{-s(\eta-\alpha)} e^{-s\alpha}|v(\beta, \alpha)|\, d\alpha$$
$$\leq \left\{ \frac{1}{4}c \int_\eta^\xi d\beta \int_0^\eta e^{-s(\eta-\alpha)}\, d\alpha + \frac{1}{2}c \int_0^\eta d\beta \int_0^\beta e^{-s(\eta-\alpha)}\, d\alpha \right\} \|v\|_s.$$

Since
$$\int_0^\eta e^{-s(\eta-\alpha)}\, d\alpha = s^{-1}(1 - e^{-s\eta}) \leq s^{-1},$$

the expression in braces does not exceed the value $s^{-1}(cT/2 + cT/2) = s^{-1}cT$, which implies that

$$\|Qv\|_s \leq cTs^{-1}\|v\|_s.$$

Setting $s_0 = 2cT$, for $s \geq s_0$ we obtain (2.22). The number T being arbitrary, we have proved the existence of the continuous function $K(x, t)$ in the infinite domain $x > t > 0$. From formulas (2.20) and (2.21) it follows that for $v \in C(\overline{\Omega})$ and $q \in C^1(\overline{\Omega})$ we have $Qv + f \in C^2(\overline{\Omega})$. Therefore, by equation (2.19), the functions v and $K(x, t)$ belong to C^2. Thus, we have proved the following statement.

Theorem 2.1. Let $a_j \in C^1[0, \infty)$. Then the operator $F = E + K$ given by formula (2.3), where the kernel $K(x, t)$ of class C^2 is determined by (2.7), (2.9) and (2.10), interlaces the operators A_1 and A_2, i.e., $Fu \in D(A_2)$ and $A_2 Fu = FA_1 u$ for all $u \in D(A_1)$.

Corollary 2.1. Suppose $A = -\partial^2 + a(x)$, $D(A) = \{u \in C^2[0, T] \mid u(0) = \alpha,\ u'(0) = 0\}$ and the function $a \in C^1[0, T]$. Then the operator A is similar to the operator $-\partial^2$ with domain of definition $D(A)$.

To prove this, it suffices to set $a_1 = a$, $a_2 = 0$, and $h_1 = h_2 = 0$ in Theorem 2.1 and to note that

$$(Fu)(0) = u(0) = \alpha.$$

Chapter 3. Inverse problems for differential equations

Before we consider the problem of determining a Sturm–Liouville operator by its spectral function, we recall some necessary information on this question. For the simplicity of presentation we restrict ourselves to the case of a finite interval. Suppose $u(x, \lambda)$ is a solution of the Cauchy problem

$$Au = -u''(x) + a(x)u = \lambda u, \qquad x \in (0, \pi), \qquad (2.23)$$

$$u(0, \lambda) = 1, \qquad u'(0, \lambda) = h. \qquad (2.24)$$

In this case the function $u(x, \lambda)$, evidently, satisfies the boundary condition

$$u'(0) - hu(0) = 0. \qquad (2.25)$$

Therefore, in order to find the eigenvalues and eigenfunctions of the boundary value problem

$$Au = \lambda u, \qquad (2.26)$$

$$u'(0) - hu(0) = 0, \qquad u'(\pi) + Hu(\pi) = 0, \qquad (2.27)$$

where h and H are given numbers, it is necessary to find the roots of the characteristic function

$$\chi(\lambda) = u'(\pi, \lambda) + Hu(\pi, \lambda) = 0. \qquad (2.28)$$

The spectrum of the boundary value problem (2.26), (2.27) is determined by these roots. Suppose a is a real function of class $C^1[0, \pi]$ and h, H are real numbers. Since the function $u_0(x, \lambda) = \cos(\sqrt{\lambda}\, x)$ is a solution of the problem (2.23)–(2.24) with $a(x) = 0$ and $h = 0$, by Theorem 2.1 there exists an operator F of the form (2.3) which interlaces the operator A with boundary condition (2.25) and the operator A_0 such that

$$A_0 = -\partial^2, \qquad \mathcal{D}(A_0) = \{u \in C^2 \mid u'(0) = 0\}.$$

In particular, $Afu_0 = FA_0u_0 = F\lambda u_0 = \lambda Fu_0$ and the function Fu_0 satisfies the boundary condition (2.25). From (2.3) it follows that $(Fu_0)(0) = u_0(0) = 1$. Thus, Fu_0 is a solution of the Cauchy problem (2.23), (2.24). By the uniqueness of the solution, we have $u(x, \lambda) = Fu_0$ or, more precisely,

$$u(x, \lambda) = \cos(\sqrt{\lambda}\, x) + \int_0^x K(x, t) \cos(\sqrt{\lambda}\, t)\, dt. \qquad (2.29)$$

From formulas (2.28) and (2.29) it follows that $\chi(\lambda)$ is an entire function. Since h, H, and $a(x)$ are real, the operator A with boundary condition (2.27) is self-adjoint and semicontinuous from below. Therefore, the zeros of $\chi(\lambda)$

are real and may be arranged in increasing order: $\lambda_1 < \lambda_2 < \ldots < \lambda_n < \ldots$.
The corresponding eigenfunctions

$$u_n(x) = u(x, \lambda_n)/\|u(x, \lambda_n)\|_{L_2(0,\pi)} \qquad (2.30)$$

form an orthonormal basis in $L_2(0, \pi)$. Let

$$\sigma(\lambda) = \sum_{\lambda_n \leq \lambda} \|u(\cdot, \lambda_n)\|^{-2}. \qquad (2.31)$$

We define the Fourier transform corresponding to the operator A by the formula

$$\Phi : f \in L_2(0, \pi) \to \hat{f}(\lambda) = \int_0^\pi f(x) u(x, \lambda) \, dx = \langle f, u(\cdot, \lambda) \rangle.$$

Since the functions $u_n(x)$ form an orthonormal basis, for all $f \in L_2(0, \pi)$ the Parseval equality holds:

$$\|f\|^2 = \sum_{n=1}^\infty |\hat{f}(\lambda)|^2 \|u(\cdot, \lambda_n)\|^{-2} = \int_{-\infty}^\infty |\hat{f}(\lambda)|^2 \, d\sigma(\lambda). \qquad (2.32)$$

Evidently, the transformation $\Phi f = \hat{f}$ maps the space $L_2(0, \pi)$ onto $L_2(\mathbb{R}; d\sigma(\lambda))$ and Φ is a unitary operator in view of (2.32).

The function $\sigma(\lambda)$ given by formula (2.31) is called *the spectral function of the operator A*. Evidently, it is uniquely defined by the spectrum $\{\lambda_n\}$ of the operator A and by the norm numbers $\varkappa_n = \|u(\cdot, \lambda_n)\|^2$. We now show that the spectral function uniquely reconstructs the numbers h, H and the function $a(x)$.

Theorem 2.2. *Suppose $\sigma_j(\lambda)$, $j = 1, 2$ is the spectral function of the boundary value problem*

$$A_j u_j = (-\partial^2 + a_j(x)) u_j = \lambda u_j, \qquad x \in (0, \pi),$$

$$u_j'(0) - h_j u_j(0) = 0, \qquad u_j'(\pi) + H_j u_j(\pi) = 0,$$

where h_j, H_j are real numbers, and $a_j(x)$ are real functions of class $C^1[0, \pi]$. If $\sigma_2(\lambda) = c\sigma_1(\lambda)$ for some $c > 0$, then $c = 1$, $h_1 = h_2$, $H_1 = H_2$, and $a_1(x) = a_2(x)$ for all $x \in [0, \pi]$.

Proof. By Theorem 2.1, the functions $u_1(x, \lambda)$ and $u_2(x, \lambda)$ are connected by the relation $u_2 = Fu_1$, where the operator $F = E + K$ is defined by formula (2.3). For an arbitrary function $f \in L_2(0, \pi)$ we set $\hat{f}_j(\lambda) = \langle f, u_j(\cdot, \lambda)\rangle$. Then, on the one hand, by the Parseval equality

$$\|f\|^2 = \int |\hat{f}_2(\lambda)|^2 \, d\sigma_2(\lambda) = c \int |\hat{f}_2(\lambda)|^2 \, d\sigma_1(\lambda), \qquad (2.33)$$

because $\sigma_2 = c\sigma_1$. On the other hand,

$$\hat{f}_2(\lambda) = \langle f, Fu_1\rangle = \langle F^*f, u_1\rangle.$$

Therefore, $\hat{f}_2(\lambda) = (F^*\hat{f}_1(\lambda))$, i.e.,

$$\|F^*f\|^2 = \int |\hat{f}_2(\lambda)|^2 \, d\sigma_1(\lambda). \qquad (2.34)$$

Comparing formulas (2.33) and (2.34), we obtain

$$\|c^{1/2}F^*f\| = \|f\|,$$

i.e., $c^{1/2}F^* = c^{1/2}E + c^{1/2}K^*$ is a unitary operator. Then K^* is a normal operator and, consequently,

$$\|(K^*)^n\|^{1/n} = \|K^*\|.$$

On the other hand, taking (2.3) into account, we see that K^* is a Volterra operator

$$(K^*f)(x) = \int_x^\pi K(t, x) f(t) \, dt$$

and therefore

$$\|(K^*)^n\|^{1/n} \to 0, \qquad n \to \infty. \qquad (2.35)$$

Hence, $K^* = 0 \Rightarrow K = 0 \Rightarrow F = E \Rightarrow c = 1$ and $u_1(x, \lambda) = u_2(x, \lambda)$ for all x and λ. This implies that $h_1 = h_2$, $H_1 = H_2$, and $a_1(x) = a_2(x)$. The theorem is proved. $\square$

Exercise 2.1. Show that each eigenvalue λ_n of the boundary value problem (2.26), (2.27) corresponds to only one eigenfunction (up to a constant multiplier). This fact means the geometric simplicity of the spectrum of the operator A.

Exercise 2.2. Prove that for real h, H, and $a(x)$ all the roots of the characteristic equation $\chi(\lambda) = 0$ are simple.

Exercise 2.3. Establish the completeness of the orthonormal system of eigenfunctions (2.30) under the assumptions of Exercise 2.2.

Exercise 2.4. Prove formula (2.35).

Exercise 2.5. Prove the analogue of Theorem 2.1 in the case where

$$\mathcal{D}(A_j) = \{u \in C^2[0, \infty) \mid u(0) = 0\}.$$

Exercise 2.6. Using Exercise 2.5, prove Theorem 2.2 replacing the boundary conditions (2.27) by the Dirichlet conditions $u(0) = u(\pi) = 0$.

Exercise 2.7. Prove Theorem 2.2 for real continuous functions by means of passing to the limit.

Exercise 2.8. Using Theorems 2.1, 2.2, and the Fourier method, investigate the problem of the uniqueness of determination of the coefficient of the boundary value problem

$$u_t = u_{xx} + a(x)u, \qquad x \in (0, \pi), \quad t > 0$$

$$u'(0, t) = 0, \qquad u'(\pi, t) = 0$$

$$u(x, 0) = f(x), \quad x \in [0, \pi] \qquad u(0, t) = g(t), \quad t \in [0, T].$$

Here the functions f and g are given and $u(x, t)$ and $a(x)$ are unknown. Consider separately the following two cases: a) f is a smooth function such that $f(x) \neq 0$ for all $x \in [0, \pi]$; b) $f(x) = \delta(x)$ is the Dirac function.

3.3. UNIQUENESS IN MULTIDIMENSIONAL INVERSE PROBLEMS IN NONSTATIONARY AND SPECTRAL STATEMENTS

We begin the investigation of uniqueness problems with the multidimensional analogue of Theorem 1.1, Chapter 2. Suppose $\Omega = \{x \in \mathbb{R}^2 \mid |x| < 1\}$

Chapter 3. Inverse problems for differential equations

and $u(x,t)$ is a solution from the class $C^\infty(\Omega \times \mathbb{R})$ of the following boundary value problem

$$iu_t + \Delta u - a(x)u = 0, \qquad x \in \Omega, \quad t \in \mathbb{R}, \tag{3.1}$$

$$\frac{\partial u}{\partial |x|}\bigg|_{\partial\Omega \times \mathbb{R}} = 0, \tag{3.2}$$

$$u(x,0) = f(x). \tag{3.3}$$

Here Δ is the Laplace operator; the potential $a(x)$ and the initial data $f(x)$ belong to the space $C^\infty(\overline{\Omega})$ and in the general case they may be complex. Given the trace of the solution $u(x,t)$ on $\partial\Omega \times [-T,T]$

$$g(x,t) = u|_{\partial\Omega \times [-T,T]} \tag{3.4}$$

it is required to reconstruct the potential $a(x)$ (f and g are known). We shall prove the uniqueness for this inverse problem by contradiction. Assume that the problem (3.1)–(3.4) has two solutions u_1, a_1, u_2, a_2 corresponding to the functions f and g. Subtract the equation for u_2, a_2 from equation (3.1), written for u_1, a_1. Denoting $u_1 - u_2 = u$, $a_1 - a_2 = a$, we obtain

$$iu_t + \Delta u - a_1 u_1 = a(x) u_2(x,t), \tag{3.5}$$

$$u(x,0) = 0, \qquad u|_\Sigma = \frac{\partial u}{\partial |x|}\bigg|_\Sigma = 0, \qquad \Sigma = \partial\Omega \times [-T,T]. \tag{3.6}$$

Set

$$P = i\partial + \Delta - a_1(x), \qquad h(x,t) = a(x)u_2(x,t).$$

Then equation (3.5) takes the form $Pu = h$. To exclude the right-hand side h, we suppose that $f(x) \neq 0$ for all $x \in \overline{\Omega}$. By continuity, we have $u_2(x,t) \neq 0$ for $x \in \overline{\Omega}$, $t \in [-T,T]$ if T is sufficiently small. For this T, the operator $Qh = (\partial_t - b(x,t))h$ is correctly defined in the cylinder $\overline{\Omega} \times [-T,T]$ (here $b(x,t) = (\partial_t u_2)/u_2$) and, evidently, $Qh = 0$. Thus, $QPu = 0$ or, what is the same,

$$PQu = [P,Q]u,$$

where $[P,Q]$ is the commutator of the operators P and Q. Introducing the new variable $v = Qu$, we have

$$Pv = [P,Q]Q^{-1}v. \tag{3.7}$$

As in Section 1, Chapter 2, the commutator $[P,Q]$ is a differential operator of the first order with respect to x and of zero order with respect to t;

Q^{-1} is a Volterra operator by the first condition of (3.6). Therefore, (3.6) and (3.7) yield

$$|iv_t + \Delta v| \le c\left(|v| + |\nabla v| + \left|\int_0^t (|v(x,\tau)| + |\nabla v(x,\tau)|)\,d\tau\right|\right)$$

$$v|_\Sigma = \left.\frac{\partial v}{\partial |x|}\right|_\Sigma = 0.$$

Repeating the arguments of Example 2.1, Chapter 2, we have $v(x,t) = 0$ for $x \in \overline{\Omega}$, $|t| \le T|x|$. Hence, $u = Q^{-1}v = 0$ and $h = Pu = 0$ in the same set; in particular, $h(x,0) = f(x)a(x) = 0$. Since $f(x) \ne 0$, we have $a(x) = 0$, i.e., $a_1 = a_2$. Thus, we have proved the following theorem.

Theorem 3.1. *Suppose u is a solution of the boundary value problem (3.1)–(3.3) and $f(x) \ne 0$ for all $x \in \overline{\Omega}$. Then the potential $a(x)$ is uniquely determined by the function $g = u|_\Sigma$, $\Sigma = \partial\Omega \times (-T,T)$.*

Now we reformulate Theorem 3.1 in spectral terms. For this purpose, we consider the problem of eigenvalues in the unit disk

$$Lu \equiv -\Delta u + a(x)u = \lambda u, \qquad x \in \Omega, \qquad (3.8)$$

$$\left.\frac{\partial u}{\partial \nu}\right|_{\partial \Omega} = 0. \qquad (3.9)$$

Here ν is the unit external normal to $\partial\Omega$; $a(x)$ is a real potential of class $C_0^\infty(\Omega)$,

$$\operatorname{supp} a \subseteq \{x \mid |x| \le s\} = \Omega_s. \qquad (3.10)$$

It is well known that the operator L with boundary condition (3.9) generates a self-adjoint operator in the Hilbert space $L_2(\Omega)$ and has purely discrete spectrum. Suppose λ_k are eigenvalues of the boundary value problem (3.8), (3.9) enumerated in increasing order and $\{u_k(x)\}$ is the corresponding orthonormal basis of eigenfunctions. We denote by $\partial\Omega'$ an arbitrary open subset of $\partial\Omega$. We call the sequence

$$S(f, \partial\Omega') = \{\lambda_k, \langle u_k, f\rangle u_k|_{\partial\Omega'}\}, \qquad k = 1, 2, \dots,$$

the spectral data of problem (3.8), (3.9). Here $\langle \cdot, \cdot \rangle$ is a scalar product in $L_2(\Omega)$ and f is a given function of class $C_0^\infty(\Omega)$ such that $f(x) \ne 0$ for $x \in \Omega_s$.

Theorem 3.2. *The spectral data $S(t, \partial\Omega')$ uniquely determine the operator L, i.e., the function a in Ω.*

Chapter 3. Inverse problems for differential equations

Proof. Let

$$u(x,t) = \sum_{k=1}^{\infty} e^{-i\lambda_k t} u_k(x) \int_{\Omega} u_k(y)\overline{f(y)}\, dy. \tag{3.11}$$

It is easy to show that $u \in C^{\infty}(\overline{\Omega} \times \mathbb{R})$, where

$$iu_t + \Delta u - a(x)u = 0, \qquad x \in \Omega, \tag{3.12}$$

$$u(x,0) = f(x), \qquad \left.\frac{\partial u}{\partial \nu}\right|_{\partial\Omega \times \mathbb{R}} = 0. \tag{3.13}$$

If we know the spectral data $S(f, \partial\Omega')$, then we know the function

$$g(x,t) = u|_{\partial\Omega \times \mathbb{R}}. \tag{3.14}$$

In the ring $\Omega \setminus \Omega_s$, we have $a(x) \equiv 0$. Therefore, equation (3.12) is transformed into an equation with constant coefficients for $x \in \Omega \setminus \Omega_s$. In this case we know the Cauchy data of its solution on $\partial\Omega' \times \mathbb{R}$. Since the surface $\partial\Omega' \times \mathbb{R}$ is not characteristic, by the Holmgren theorem, we know the function $u(x,t)$ in the whole cylindrical layer $(\Omega \setminus \Omega_s) \times \mathbb{R}$ and, by continuity, we know u and $\partial u/\partial|x|$ on $\partial\Omega_s \times \mathbb{R}$. Application of Theorem 3.1 with Ω replaced by Ω_s finishes the proof. □

We construct the function

$$\Theta(x,y,\lambda) = \sum_{\lambda_k < \lambda} u_k(x) u_k(y), \qquad x, y \in \overline{\Omega}, \quad \lambda \in \mathbb{R}^1$$

which is called *the spectral function of the operator L*, and set

$$\rho(x,\lambda) = \int \Theta(x,y,\lambda) f(y)\, dy.$$

In terms of the function $\rho(x,\lambda)$, equalities (3.11) and (3.14) become as follows:

$$u(x,t) = \int e^{-i\lambda t}\, d\rho(x,\lambda), \qquad x \in \overline{\Omega},\ t \in \mathbb{R},$$

$$g(x,t) = \int e^{-i\lambda t}\, d\rho(x,\lambda), \qquad x \in \Omega',\ t \in \mathbb{R}.$$

Therefore, the proof of Theorem 3.2 implies the following corollary.

Corollary 3.1. The operator L is uniquely determined by the measure $d\rho(x,\lambda)|_{\partial\Omega'\times\mathbb{R}}$.

Since the solution of the boundary value problem for the parabolic equation

$$v_t = \Delta v - a(x)v, \qquad x \in \overline{\Omega}, \quad t \in [0,T], \tag{3.15}$$

$$v(x,0) = f(x), \qquad x \in \Omega, \tag{3.16}$$

$$\left.\frac{\partial u}{\partial \nu}\right|_{\partial\Omega\times[0,T]} = 0 \tag{3.17}$$

is the Laplace transform of the measure $d\rho(x,\lambda)$:

$$v(x,t) = \int e^{-\lambda t}\, d\rho(x,\lambda), \qquad t \in [0,T], \tag{3.18}$$

taking into account the uniqueness of the solution of equation (3.18), the measure $d\rho(x,\lambda)$, $x \in \partial\Omega'$, $\lambda \in \mathbb{R}$ is uniquely determined by the function $v|_\Sigma = \partial\Omega' \times [0,T]$. Thus, from Corollary 3.1 we obtain the following theorem.

Theorem 3.3. Suppose v is a solution of the boundary value problem (3.15)–(3.17). Then the coefficient $a(x)$ is uniquely determined by the function $v|_\Sigma$, $\Sigma = \partial\Omega' \times [0,T]$.

Analogously, if $w(x,t)$ is a solution of the boundary value problem for the hyperbolic equation

$$w_{tt} = \Delta w - a(x)w, \qquad x \in \Omega, \quad t > 0, \tag{3.19}$$

$$w(x,0) = f(x), \qquad w_t(x,0) = 0, \quad x \in \Omega, \tag{3.20}$$

$$\left.\frac{\partial w}{\partial \nu}\right|_{\partial\Omega\times[0,\infty)} = 0, \tag{3.21}$$

then we have

$$w(x,t) = \int \cos(\sqrt{\lambda}\,t)\, d\rho(x,\lambda)$$

for $a(x) \geq 0$. This means that $w(x,t)$ is the cosine-transformation of the measure $d\rho(x,\lambda)$. As a result, we obtain the following theorem.

Theorem 3.4. Suppose w is a solution of the boundary value problem (3.19)–(3.21) and $a(x) \geq 0$ for all $x \in \Omega$. Then the coefficient $a(x)$ is uniquely determined by the function $w|_\Sigma$, $\Sigma = \partial\Omega' \times [0,\infty)$.

Arguing as in the proofs of Theorems 3.1–3.4 we prove the theorems on the uniqueness of determination of some coefficients (or all coefficients)

Chapter 3. Inverse problems for differential equations

of the general elliptic operator L by its spectral data or by the traces of solutions of nonstationary equations

$$i\partial_t u = Lu, \qquad v_{tt} = Lv,$$

and so on. Some of these equations are mentioned below.

Exercise 3.1. Replace the Neumann boundary condition in Theorem 3.2 by $\partial u/\partial \nu + \delta(x)u(0)$, $x \in \partial\Omega$, where $\sigma(x) \geq 0$, $\sigma \in C^\infty(\partial\Omega)$, and $\sigma(x) = 0$ for $x \in \partial\Omega'$. Prove that the spectral data $S(f, \partial\Omega')$ uniquely determine the potential $a(x)$ and the function $\sigma(x)$.

Exercise 3.2. Generalize Theorems 3.1–3.4 to the n-dimensional case.

Exercise 3.3. Suppose $\Omega = \{x \in \mathbb{R}^n \mid |x| < 1\}$ and for all multi-indices α such that $|\alpha| \leq 2m - 1$, $m = [n/4] + 1$ the functions $a_\alpha = \partial^\alpha a|_{\partial\Omega}$ are known. Then, if a given function $f \in C^\infty(\overline{\Omega})$ is such that $f(x) \neq 0$ for all $x \in \overline{\Omega}$ and the functions $L^p f$ satisfy the Neumann condition for $p = 0, 1, \ldots, m$, then the spectral data $S(f, \partial\Omega)$ uniquely determine the operator $L = -\Delta + a(x)$.

Exercise 3.4. Consider the self-adjoint operator L generated by the differential expression

$$ly \equiv \sum_{m=0}^{n}(a_{n-m}(x)y^{(m)})^m, \qquad x \in (-1, 1)$$

and the boundary conditions

$$y(\pm 1) = y'(\pm 1) = \ldots = y^{(m-1)}(\pm 1) = 0.$$

We assume that real coefficients satisfy the conditions $a_m \in C^\infty[-1, 1]$, $m = 0, 1, \ldots, n$; $a_0(x) \geq c > 0$, where c is independent of x, $\operatorname{supp} a_m \subseteq [-s, s]$ for $m = 1, 2, \ldots, n$, $a_0(x) = 1$ for $x \in [-1, 1] \setminus [-s, s]$, $s < 1$. Suppose λ_k are eigenvalues of the operator L and u_k are the corresponding eigenfunctions. Let

$$S_f = \{\lambda_k, \langle u_k, f\rangle u_k^m(-1)\}, \qquad m = n, n+1, \ldots, 2n-1, \quad k = 1, 2, \ldots,$$

where $f \in C_0^\infty(-1, 1)$. Prove that the sequence S_f uniquely determines the coefficient a_{n-p} of the operator L if the other coefficients are known and $f^{(p)}(x) \neq 0$ for all $x \in [-s, s]$.

Exercise 3.5. For a prescribed set of functions $f_j(x)$ in Exercise 3.4, find the conditions under which the sequences S_{f_j} uniquely determine all the coefficients of the operator L.

Exercise 3.6. Suppose that in the half-plane $x_n > 0$, $t \in \mathbb{R}$ the function $u(x,t)$ is a solution of the boundary value problem

$$iu_t + c^2(x)\rho(x)\sum_{j=1}^{n}\partial_j(\rho^{-1}(x)\partial_j u) = 0,$$

$$u(x,0) = f(x), \qquad \partial_n u|_{x_n=0} = 0, \qquad u|_{x_n=0} = g(x',t).$$

Here $\partial_j = \partial/\partial x$, $c(x), \rho(x) \geq c_0 > 0$, $c, \rho \in C^\infty(x_n \geq 0)$. Assuming that one of the functions ρ or c is known, find the sufficient conditions for the uniqueness of determination of the other function by the given f and g.

Exercise 3.7. Setting two different initial conditions $f_j(x)$, $j = 1, 2$, in Exercise 3.6, find the conditions under which both coefficients $c(x)$ and $\rho(x)$ are uniquely determined by f_j and g_j.

Exercise 3.8. Reformulate Exercises 3.6 and 3.7 in terms of the measure $d\rho(x, \lambda)$ and prove the corresponding analogues of Theorems 3.3, 3.4.

Exercise 3.9. State and consider the problem of determining the operator

$$L = \sum_{i,j=1}^{n}\partial_j(a_{ij}(x)\partial_j), \qquad |x| < 1$$

with the Dirichlet or Neumann boundary conditions on $\partial\Omega$ by the spectral data $S(f_k, \partial\Omega)$, $k = 1, 2, \ldots, M$.

3.4. REMARKS AND REFERENCES

To Section 1. The inversion formula (1.25) was obtained by Dobrinskii and Lavrent'ev (1978). The uniqueness theorem in exact statement was first proved by Romanov (1973).

To Section 2. Transformation operators were introduced by Delsarte (1938) in the context of the theory of generalized shift operators. The interlacing operator F in the general case was constructed by Povzner (1948).

Theorem 2.2 was first proved by Marchenko (1950). The idea of the proof of Theorem 2.1 based on the introduction of a norm with a weight in the space C is contained in the monograph by Krasnoselskii and Zabreiko (1975). This idea is similar to the Carleman method used in Chapter 2. Detailed investigation of inverse problems for the Sturm–Liouville operators can be found in the monographs by Marchenko (1972, 1977) and Levitan (1962, 1984), the ideas of which were used in this section. Other references can also be found there. Extensive bibliography is presented in Shadon and Sabatier (1980).

To Section 3. This section is based on the works of the author (Bukhgeim, 1984, 1985). Some related inverse problems in nonstationary statement were investigated by Klibanov (1984, 1985). The problem of substantiation of the Fourier method for the solution of mixed problems is considered in Ladyzhenskaya (1953) and Maurin (1959).

Chapter 4.

Volterra operator equations and their applications

4.1. VOLTERRA OPERATOR EQUATIONS IN SCALES OF BANACH SPACES

In this section we study operator equations of the form

$$u = Vu + f$$

where the operator V has a Volterra majorant in a certain sense.

Consider an interval $I = (a, b)$ and a one-parameter set of Banach spaces X_s, $s \in I$, with norm $\|\cdot\|_s$ such that for $s' < s$

$$X_s \subseteq X_{s'}, \qquad \|u\|_{s'} \leq \|u\|_s, \quad \forall u \in X_s. \tag{1.1}$$

Note that the linear space X_s is a subspace of the linear space $X_{s'}$.

Then the linear space $\mathcal{X} = \cup_{s \in I} X_s$ is said to be *a scale of Banach spaces*. The scale is called *continuous* if the equality $\lim_{s' \to s} \|u\|_{s'} = \|u\|_s$ holds for any $s', s \in I$, $s' < s$, $u \in X_s$.

We denote by $\mathcal{L}(\mathcal{X})$ the set of linear operators $A : \mathcal{X} \to \mathcal{X}$ such that $A \in \mathcal{L}(X_s, X_{s'})$ for any $s, s' \in I$, $s' < s$. Evidently, $\mathcal{L}(\mathcal{X})$ is an algebra. In particular, if $A \in \mathcal{L}(\mathcal{X})$, then $A^n \in \mathcal{L}(\mathcal{X})$ for any integer n. For $s' < s$ and $A \in \mathcal{L}(\mathcal{X})$ we set

$$r(s, s', A) = \varlimsup_{n \to \infty} \|A^n\|_{s,s'}^{1/n}, \tag{1.2}$$

where $\|A\|_{s,s'} = \|A\|_{\mathcal{L}(X_s, X_{s'})}$. The number r is an analogue of the spectral radius of a bounded operator. For a similar case, the following lemma holds.

Lemma 1.1. Suppose $A \in \mathcal{L}(X)$ and for some $s, s' \in I$, $s' < s$ the inequality $r(s, s', A) < 1$ holds. Then for any $f \in X_s$ the equation

$$u = Au + f$$

has a solution $u \in X_{s'}$ such that

$$\|u\|_{s'} \leq c(s, s')\|f\|_s.$$

This solution is unique in the space X_s.

Proof. Let

$$u = \sum_{n=0}^{\infty} A^n f. \tag{1.3}$$

Since $r(s, s', A) < 1$, by the Cauchy criterion, the series (1.3) converges in the norm of the space $X_{s'}$ to a certain element u such that

$$\|u\|_{s'} \leq c\|f\|_s,$$

$$c(s, s') = \sum_{k=0}^{n} \|A^k\|_{s,s'}. \tag{1.4}$$

The fact that u is a solution of the equation $u = Au + f$ is evident. Suppose that $f = 0$, $u \in X_s$. Then for any integer n we have $u = A^n u$ and $\|u\|_{s'} \leq \|A^n\|_{s,s'}\|u\|_s$. By the convergence of the series (1.4), we have $\|A^n\|_{s,s'} \to 0$ for $n \to \infty$. Hence, $\|u\|_{s'} = 0$ and therefore $u = 0$. The lemma is proved. $\square$

We shall give some examples of scales of functional Banach spaces and the estimates of differentiation operators and operators of multiplication by a function in these scales. These estimates are used for derivation of the number $r(s, s', A)$ in applications.

Example 1.1. Let $\mathbb{C}^n$ be the n-dimensional complex space and Ω_s, $s \in I = (a, b)$, be a collection of open sets in $\mathbb{C}^n$ such that $\Omega_{s'} \subset \Omega_s$ for $s' < s$ and the distance between the boundary $\partial\Omega_s$ and $\Omega_{s'}$ is equal to $r(s - s')$. We denote by X_s the set of functions $u(z)$ analytic and bounded in the domain Ω_s with norm

$$\|u\|_s = \sup_{\Omega_s} |u(z)|. \tag{1.5}$$

Chapter 4. Volterra operator equations and their applications

By well-known theorems of complex analysis, X_s is a Banach space. Evidently, for $s' \leq s$ we have

$$X_s \subseteq X_{s'}, \qquad \|u\|_{s'} \leq \|u\|_s.$$

Thus, $\mathcal{X} = \cup X_s$, $s \in I$ is a scale of Banach spaces. For $a, u \in X_s$ we have $\|au\|_s \leq \|a\|_s \|u\|_s$. Therefore, the operator of multiplication by the function $a \in X_s$ is bounded. Let $\partial_j = \partial/\partial z_j$. We show that

$$\|\partial_j u\|_{s'} \leq (r(s - s'))^{-1} \|u\|_s. \qquad (1.6)$$

First, we suppose that $n = 1$. Since for any point $z \in \Omega_{s'}$ the disk of radius $r(s - s')$ lies in Ω_s by assumption, from the Cauchy formula it follows that

$$u'(z) = \frac{1}{2\pi i} \int_{|z-\xi|=r(s-s')} \frac{u(\xi)}{(\xi - z)^2} d\xi, \qquad u \in X, \quad z \in \Omega_{s'}. \qquad (1.7)$$

Hence, passing to the polar coordinates

$$\xi = z + \rho \exp(i\varphi), \qquad \rho = r(s - s'), \qquad \varphi \in [0, 2\pi],$$

we obtain

$$|u'(z)| \leq (r(s - s'))^{-1} \|u\|_s, \qquad z \in \Omega_{s'},$$

or

$$\|\partial u\|_{s'} \leq (r(s - s'))^{-1} \|u\|_s.$$

In the case of n variables, using formula (1.7) with respect to the variable z_j, we obtain estimate (1.6). To estimate the derivative $\partial^\alpha = \partial_1^{\alpha_1} \partial_2^{\alpha_2} \ldots \partial_n^{\alpha_n}$ we divide the segment $[s', s]$ into $|\alpha| = \alpha_1 + \alpha_2 + \ldots + \alpha_n$ parts of length $(s-s')/|\alpha|$. Using formula (1.6) for successive evaluation of the derivatives ∂_k from $X_{s_{j+1}}$ to X_{s_j}, ($s_0 = s'$, $s_{j+1} = s_j + (s - s')/|\alpha|$, $j = 0, \ldots, |\alpha| - 1$), we obtain

$$\|\partial^\alpha u\|_{s'} \leq (|\alpha|/(r(s - s')))^{|\alpha|} \|u\|_s. \qquad (1.8)$$

Example 1.2. Suppose $\lambda(\xi)$ is a nonnegative function and $\hat{u}(\xi)$ is the Fourier transform of the function $u(x)$:

$$\hat{u}(\xi) = \int \exp(-i \langle x, \xi \rangle) u(x) \, dx.$$

We denote by H_s the Hilbert space obtained by taking closure in the norm

$$\|u\|_s^2 = \int \exp(2s\lambda(\xi)) |\hat{u}(\xi)|^2 \, d\xi, \qquad s \in (-\infty, \infty)$$

of the set of functions u such that $\hat{u} \in C_0^\infty(\mathbb{R}^n)$. Evidently, the set H_s, $s \in I = (-\infty, \infty)$, forms a scale of Hilbert spaces.

Proposition 1.1. Suppose that a function λ satisfies the condition

$$0 \leq \lambda(\xi + \zeta) \leq \lambda(\xi) + \lambda(\zeta), \qquad \xi, \zeta \in \mathbb{R}^n, \tag{1.9}$$

and a function $a(x)$ is such that $\exp(|s|\lambda)\hat{a} \in L_1(\mathbb{R}^n)$. Then

$$\|au\|_s \leq (2\pi)^{-n} \|\exp(|s|\lambda)\hat{a}\|_{L_1} \|u\|_s. \tag{1.10}$$

Proof. Since $\widehat{au} = (2\pi)^{-n} \hat{a} * \hat{u}$, we have

$$e^{s\lambda(\xi)} \widehat{au}(\xi) = \frac{1}{(2\pi)^n} \int e^{s\lambda(\xi) - s\lambda(\zeta) - |s|\lambda(\xi - \zeta)} e^{|s|\lambda(\xi - \zeta)} \hat{a}(\xi - \zeta) e^{s\lambda(\zeta)} \hat{u}(\zeta) \, d\zeta$$

$$\leq \frac{1}{(2\pi)^n} \int e^{|s|\lambda(\xi - \zeta)} \hat{a}(\xi - \zeta) e^{s\lambda(\zeta)} \hat{u}(\zeta) \, d\zeta$$

$$= \frac{1}{(2\pi)^n} (e^{|s|\lambda} \hat{a}) * (e^{s\lambda} \hat{u}).$$

Here we have used the inequality $s(\lambda(\xi) - \lambda(\zeta)) \leq |s| \lambda(\xi - \zeta)$ which holds by (1.9). Applying the Young inequality $\|\varphi * v\|_{L_p} \leq \|\varphi\|_{L_1} \|v\|_{L_p}$ for $p = 2$, we obtain (1.10). The proposition is proved. $\square$

Now we estimate the derivatives. Let

$$N(s, s') = \sup_\xi (|\xi_j| \exp(-\lambda(\xi)(s - s'))). \tag{1.11}$$

Since $\partial_j \hat{u} = i\xi_j \hat{u}(\xi)$, $i^2 = -1$, we have

$$\|\partial_j u\|_{s'}^2 = \int \exp(2s\lambda) \exp(-2(s - s')\lambda) \xi_j^2 |\hat{u}(\xi)|^2 \, d\xi \leq N^2(s, s') \|u\|_s^2.$$

Thus,

$$\|\partial_j u\|_{s'} \leq N(s, s') \|u\|_s.$$

In particular, for $\lambda(\xi) = |\xi|$, from (1.11) we obtain

$$N(s, s') = e^{-1}(s - s')^{-1}. \tag{1.12}$$

If $n = 1$, for the multiplication operator to be bounded in H_s, $s > 0$, it suffices to require that the function $a(x + iy)$ be analytic in the strip $|y| < s$, $\bar{a} = \sup |a(x + iy)| < \infty$. In this case

$$\|au\|_s \leq \sqrt{2}\, \bar{a} \|u\|_s. \tag{1.13}$$

Indeed, since $\exp(\mp s\xi)\widehat{au}(\xi)$ coincides with the Fourier transform of the function $a(x \pm is)u(x \pm is)$, using the Parseval equality and the trivial estimate $e^{2s\xi} \leq e^{2s\xi} + e^{-2s\xi} \leq 2e^{2s|\xi|}$, we obtain (1.13).

The space H_{-s}, $s > 0$, is conjugate to H_s with respect to the scalar product in $L_2 \equiv H_0$.

On the basis of a given scale $\mathcal{X}$ of Banach spaces, we can construct scales of functions with values in $\mathcal{X}$ using the following simple procedure.

Suppose Ω is an open set in $\mathbb{R}^n$. Given the Banach space $L_p(\Omega)$, $p \geq 1$, and a scale $\mathcal{X}$, it is natural to construct the scale $\mathcal{Y} = L_p(\Omega; \mathcal{X})$ consisting of functions $u : \Omega \to \mathcal{X}$ such that

$$\mathcal{Y} = \cup_{s \in I} Y_s, \qquad Y_s = L_p(\Omega; X_s),$$

$$\|u\|_{Y_s} = \|u\|_{p,s} = \left\{ \int_\Omega \|u(t)\|_s^p \, dt \right\}^{1/p}, \qquad 1 \leq p < \infty,$$

$$\|u\|_{\infty,s} = \operatorname*{vrai\,sup}_{t \in \Omega} \|u(t)\|_s.$$

The scale

$$C(\overline{\Omega}; \mathcal{X}) = \cup_{s \in I} C(\overline{\Omega}; X_s),$$

$$\|u\|_{C(\overline{\Omega}; X_s)} = \sup_{t \in \Omega} \|u(t)\|_s,$$

is defined analogously. Here $C(\overline{\Omega}; X_s)$ is the Banach space of functions $u : \Omega \to X_s$ that are continuous and bounded in the closure $\overline{\Omega}$ of the set Ω.

Suppose A is a linear bounded operator acting in $L_p(\Omega)$ such that $\operatorname{Sp} A = \{0\}$ and $(Au)(t) \geq 0$ for real nonnegative functions $u(t)$.

Definition 1.1. An operator V from $\mathcal{L}(L_p(\Omega; \mathcal{X}))$ belongs to the class $V_p(\mathcal{X}, A, \alpha)$, $\alpha \geq 0$ if for all $s, s' \in I$, $s' < s$, $u \in L_p(\Omega; X_s)$ and for almost all $t \in \Omega$ the following estimate holds:

$$\|(Vu)(t)\|_{s'} \leq (s - s')^{-\alpha} (A\|u\|_s)(t). \tag{1.14}$$

For any linear bounded operator A with $\operatorname{Sp} A = \{0\}$ which acts in a Banach space, we define by $\rho(A)$ and $\sigma(A)$ the order and the type of the entire function

$$f_A(\lambda) = \sum_{k=0}^{n} \|A^k\| \lambda^k, \qquad \lambda \in \mathbb{C}, \tag{1.15}$$

respectively.

Using the well-known formula which expresses the order and type of an entire function in terms of the Taylor coefficients, we have

$$\rho(A) = -\varlimsup_{n\to\infty}(n \ln n / \ln \|A^n\|), \qquad (1.16)$$

$$\sigma(A) = (e\rho(A))^{-1}(\varlimsup_{n\to\infty} n^{1/\rho(a)} \|A^n\|^{1/n})^{\rho(A)}. \qquad (1.17)$$

Since

$$\operatorname{Sp} A = 0 \quad \Leftrightarrow \quad \varlimsup_{n\to\infty} \|A^n\|^{1/n} = 0, \qquad (1.18)$$

from formulas (1.16) and (1.17) it follows that $\rho(A) \geq 0$, $\sigma(A) \geq 0$. For $\rho(A) = \infty$ we put $\sigma(A) = \infty$. The operator A is called *quasinilpotent* if it satisfies the second condition of (1.19).

Now we are ready to formulate the theorem on the solvability of the equation

$$u = Vu + f, \qquad V \in \mathcal{V}_p(\mathcal{X}, A, \alpha). \qquad (1.19)$$

Theorem 1.1. *Suppose that $V \in \mathcal{V}_p(\mathcal{X}, A, \alpha)$, $f \in L_p(\Omega; X_s)$. Then, if $\alpha\rho(A) < 1$ or $\alpha\rho(A) = 1$, $e\rho(A)\sigma(A) < s - s'$, then equation (1.19) has a solution $u \in L_p(\Omega; X_{s'})$, $\|u\|_{p,s'} \leq c\|f\|_{p,s}$, where the number c is independent of u and f. This solution is unique in $L_p(\Omega; X_s)$.*

Proof. For a sequence $s_j \in I$ such that $s_0 = s' < s$, $s_{j+1} = s_j + (s-s')/n$, $j = 0,\ldots,n-1$ from estimate (1.5) we obtain

$$\|V^n u\|_{s'} \leq (n/(s-s'))^{\alpha n} A^n \|u\|_s$$

by induction. Hence, from the definition of the numbers $r(s, s', V)$, $\rho(A)$, and $\sigma(A)$ we have

$$r(s, s', V) \leq [e\rho(A)\sigma(A)/(s-s')]^{1/\rho}$$

for $\alpha\rho(A) = 1$ and $r(s, s', V) = 0$ for $\alpha\rho(A) < 1$. Application of Lemma 1.1 finishes the proof. $\square$

In some applications it is important to have the estimates of the norm of solution of equation (1.19) by the norm of the right-hand side f in the same space. We formulate the corresponding theorem restricting ourselves for the sake of simplicity to the case of the Hilbert scale $\mathcal{X} = \cup X_s$, $s \in I = (-a,a)$ of the form

$$\|u\|_s = \|\exp(s\Lambda^{1/\alpha})u\|, \qquad s \in I.$$

Chapter 4. Volterra operator equations and their applications

Here $X = X_0$ is a separable Hilbert space with norm $\|\cdot\| = \|\cdot\|_0$, Λ is a self-adjoint nonnegative operator in X with domain of definition $\mathcal{D}(\Lambda)$ which is dense everywhere in X. Thus, the Hilbert space X_s, $s \in I$, is defined as the closure in the norm $\|u\|_s$ of the set of elements $u \in X$ such that $\|u\|_s < \infty$.

A typical case of such a scale was considered in Example 1.2, where $X = L_2(\mathbb{R}^n)$, $\alpha = 1$, and the operator Λ is defined by the formula

$$\widehat{\Lambda u}(\xi) = \lambda(\xi)\hat{u}(\xi)$$

where $u \in \mathcal{D}(\Lambda) \Leftrightarrow \Lambda\hat{u} \in L_2(\mathbb{R}^n)$. Usually $\lambda(\xi) = |\xi|$.

Theorem 1.2. Let $V \in \mathcal{V}_p(\mathcal{X}, A, \alpha)$. Assume that either $\alpha\rho(A) < 1$ or $\alpha\rho(A) = 1$, $e\sigma(A)\rho(A) < s$. Then a solution of the equation $u = Vu + f$ is unique in the space $L_p(\Omega; X)$ and is densely solvable in it. If $u \in M = \{u \mid \|\Lambda u\|_{p,0} \leq m\}$, $f \in L_p(\Omega; X)$, then the following stability estimate holds:

$$\|u\|_{p,0} \leq \omega_m(\|f\|_{p,0}), \tag{1.20}$$

$$\omega_m(\delta) \sim m(s/\ln\delta^{-1})^\alpha, \qquad \delta \to 0. \tag{1.21}$$

Proof. The uniqueness of the solution of equation (1.9) follows from Theorem 1.1 for $s = 0$. To prove the dense solvability, we note that by Theorem 1.1 there exists a solution $u \in L_p(\Omega; X_0) = L_p(\Omega; X)$ for any function $f \in L_p(\Omega; X_s)$, where $s > 0$ for $\alpha\rho(A) < 1$ and $s > e\sigma(A)\rho(A)$ for $\alpha\rho(A) = 1$. Since the space $L_p(\Omega; X_s)$ is dense in $L_p(\Omega; X)$ (because X_s is dense in X), the dense solvability is established. Now, we prove estimate (1.20). By Theorem 1.1, for $f \in L_p(\Omega; X)$ there exists a solution $u \in L_p(\Omega; X_{-s})$ such that

$$\|u\|_{p,-s} \leq c\|f\|_{p,0}. \tag{1.22}$$

For any $\varepsilon > 0$ the following interpolation estimate holds:

$$\|u\|_0 \leq \varepsilon\|\Lambda u\|_0 + \exp(s\varepsilon^{-1/\alpha})\|u\|_{-s} \tag{1.23}$$

Therefore, by the Minkowski inequality we obtain

$$\|u\|_{p,0} \leq \varepsilon\|\Lambda u\|_{p,0} + \exp(s\varepsilon^{-1/\alpha})\|u\|_{p,-s}.$$

In view of estimate (1.22), the foregoing inequality implies

$$\|u\|_{p,0} \leq \varepsilon\|\Lambda u\|_{p,0} + c\exp(s\varepsilon^{-1/\alpha})\|f\|_{p,0}. \tag{1.24}$$

Suppose $\|\Lambda u\|_{p,0} \le m$ and $\|f\|_{p,0} \le \delta$. Then
$$\|u\|_{p,0} \le \varepsilon m + c\exp(s\varepsilon^{-1/\alpha})\delta.$$
Since the left-hand side of the inequality is independent of ε, we have
$$\|u\|_{p,0} \le \omega_m(\delta), \qquad \omega_m(\delta) = m \inf_{e>0} \varphi(\varepsilon),$$
$$\varphi(\varepsilon) = \varepsilon + c\exp(s\varepsilon^{-1/\alpha})\delta/m.$$
The function $\varphi'(\varepsilon)$ vanishes at the minimum point $\varepsilon = \varepsilon(\delta)$, i.e.,
$$(1/\varepsilon)^{1+\alpha^{-1}} \exp(s\varepsilon^{-1/\alpha}) = \alpha m/cs\delta. \qquad (1.25)$$
This formula yields
$$s\varepsilon^{-1/\alpha}(\delta) \sim \ln\delta^{-1}, \qquad \delta \to 0$$
or, putting it another way,
$$\varepsilon(\delta) \sim (s/\ln\delta^{-1})^\alpha, \qquad \delta \to 0.$$
Taking into account (1.25), we have
$$\varphi(\varepsilon(\delta)) = \varepsilon(\delta) + \alpha[\varepsilon(\delta)]^{1+\alpha^{-1}}/s \sim \varepsilon(\delta),$$
and we obtain estimate (1.20), (1.21).

It remains to prove estimate (1.23) used above. To this end, we recall the spectral theorem: There exists an isometric mapping $F : u \to \hat{u}$ of the space X onto the direct integral
$$\hat{X} = \int_0^\infty \hat{X}(\lambda)\,d\mu(\lambda)$$
of Hilbert spaces which diagonalizes the operator Λ:
$$(\hat{\Lambda} u)(\lambda) = \lambda \hat{u}(\lambda).$$
Here $\hat{u} = \{\hat{u}(\lambda)\}$ is the image of the element u under the mapping F,
$$\|u\|^2 = \int_0^\infty \|\hat{u}(\lambda)\|^2_{\hat{X}(\lambda)}\,d\mu(\lambda); \qquad (1.26)$$
the measure $d\mu(\lambda)$ is concentrated in the interval $[0,\infty)$ since $\Lambda \ge 0$. (The proof of this theorem is given, for example, in Maurin (1959)).

Chapter 4. Volterra operator equations and their applications

We now write equation (1.26) in the form

$$\|u\|^2 = \int_{\lambda<1/\varepsilon} \|\hat{u}(\lambda)\|^2_{\hat{X}(\lambda)} \, d\mu(\lambda) + \int_{\lambda \geq 1/\varepsilon} \|\hat{u}(\lambda)\|^2_{\hat{X}(\lambda)} \, d\mu(\lambda) = I_1 + I_2$$

and evaluate the integrals I_1 and I_2 separately:

$$I_1 = \int_{\lambda<1/\varepsilon} \exp(2s\lambda^{1/\alpha}) \exp(-2s\lambda^{1/\alpha}) \|\hat{u}(\lambda)\|^2_{\hat{X}(\lambda)} \, d\mu(\lambda)$$

$$\leq \exp(2s\varepsilon^{-1/\alpha}) \|\exp(-s\Lambda^{1/\alpha})u\|^2 = \exp(2s\varepsilon^{-1/\alpha}) \|u\|^2_{-s},$$

$$I_2 = \int_{\lambda \geq 1/\varepsilon} \lambda^{-2} \|\lambda \hat{u}(\lambda)\|^2_{\hat{X}(\lambda)} \, d\mu(\lambda) \leq \varepsilon^2 \|\Lambda u\|^2.$$

Substituting these estimates into (1.26) and extracting the square root, we obtain (1.22). The theorem is proved. □

In order to use Theorems 1.1 and 1.2, it is necessary to calculate or estimate from above the numbers $\rho(A)$ and $\sigma(A)$. Together with formulas (1.16) and (1.17), the following two lemmas will be useful.

Lemma 1.2. *Suppose X is a complex Banach space and A is a quasinilpotent operator acting in X. We set*

$$M_A(r) = \sup \|(E - \lambda A)^{-1}\|, \qquad \lambda \in \mathbb{C}, \quad |\lambda| = r.$$

Then

$$\rho(A) = \overline{\lim_{r \to \infty}} \, (\ln \ln M_A(r) / \ln r), \tag{1.27}$$

$$\sigma(A) = \overline{\lim_{r \to \infty}} \, (\ln M_A(r) / r^{\rho(A)}). \tag{1.28}$$

Lemma 1.3. *(i) For any complex number α and integer $k \geq 1$, we have*

$$\rho(\alpha A) = \rho(A), \qquad \sigma(\alpha A) = |\alpha|^{\rho(A)} \sigma(A)$$

$$\rho(A^k) = \rho(A)/k, \qquad \sigma(A^k) = \sigma(A).$$

(ii) If F is a linear isomorphism from a Banach space X' onto X and $A' = F^{-1}AF$, then

$$\rho(A') = \rho(A), \qquad \rho(A') = \rho(A).$$

The proof of Lemma 1.3 follows immediately from formulas (1.16), (1.17), (1.27), and (1.28) and can be established by the reader.

Proof of Lemma 1.2. To prove Lemma 1.2, for the time being we denote the numbers defined by formulas (1.27) and (1.28) by $\rho'(A)$ and $\sigma'(A)$, respectively, and show that $\rho(A) = \rho'(A)$ and $\delta(A) = \delta'(A)$. Since $M_A(r) \leq M_f(r)$, where $M_f(r) = \sup_{|\lambda|=r}(f_A(\lambda))$ and the function $F_A(\lambda)$ is defined by (1.15), then $\rho'(A) \leq \rho(A)$. Moreover, if $\rho'(A) = \rho(A)$, then $\sigma'(A) \leq \sigma(A)$. In particular, if $\rho'(A) = \infty$ or $\sigma'(A) = \infty$ then $\rho(A) = \infty$ or $\sigma(A) = \infty$, correspondingly.

Let us obtain the inverse inequalities assuming that $\rho'(A) < \infty$ and $\sigma'(A) < \infty$. By the Cauchy formula, we have

$$A^n = \frac{1}{2\pi i} \int_\gamma \lambda^n (\lambda - A)^{-1} d\lambda,$$

where γ is an arbitrary smooth contour which does not pass through the origin. Making the change of variables $\lambda^{-1} = \mu$, we obtain

$$A^n = \frac{1}{2\pi i} \int_{\gamma'} \mu^{-n-1}(E - \mu A)^{-1} d\mu, \qquad (1.29)$$

where γ' is the image of γ for the inversion $1/\lambda = \mu$.

For any $\varepsilon > 0$, by the definition of ρ' and σ', there exists $r(\varepsilon)$ such that

$$\|(E - \mu A)^{-1}\| \leq \exp((\sigma' + \varepsilon) r^{\rho'}), \qquad |\mu| = r > r(\varepsilon).$$

Hence, taking the circle $|\mu| = r$ for γ', from (1.29) we obtain

$$\|A^n\| \leq r^{-n} \exp((\sigma' + \varepsilon) r^{\rho'}). \qquad (1.30)$$

The right-hand side of (1.30) is minimal for $r = (n/((\sigma' + \varepsilon)\rho'))^{1/\rho'}$. Therefore,

$$\|A^n\| \leq e^{n/\rho'} ((\sigma' + \varepsilon) \rho'/n)^{n/\rho'}. \qquad (1.31)$$

From this inequality it follows that $\rho(A) \leq \rho'(A)$. Since we have established the inverse inequality, it follows that $\rho(A) = \rho'(A)$, $\sigma(A) \leq \sigma'(A) + \varepsilon$. Since ε is arbitrary, we obtain $\sigma(A) = \sigma'(A)$. Lemma 1.2 is proved. □

Lemma 1.2 shows that the numbers $\rho(A)$ and $\sigma(A)$ are the order and the type of the entire operator function $(E - \lambda A)^{-1}$, respectively. Further, for brevity, we shall call them *the order* and *the type of the operator A*. The most important example of a quasinilpotent operator A is the classical Volterra operator acting in the Banach space $L_p(0,T)$ or in $C[0,T]$:

$$(Au)(t) = \int_0^t A(t,\tau) u(\tau) d\tau, \qquad t \in [0,T], \qquad (1.32)$$

with sufficiently smooth kernel $A(t,\tau)$. For $A(t,\tau) \equiv 1$ it becomes the integration operator J:

$$(Ju)(t) = \int_0^t u(\tau)\,d\tau. \qquad (1.33)$$

We show that
$$(1+np)^{-1/p} T^n/n! \leq \|J^n\| \leq T^n/n!. \qquad (1.34)$$

Indeed, from formula (1.33), by induction we obtain

$$(J^n u)(t) = \frac{1}{(n-1)!} \int_0^t (t-\tau)^{n-1} u(\tau)\,d\tau. \qquad (1.35)$$

Therefore, $J^n u$ is the convolution $k * u$, where

$$k(t) = \begin{cases} t^{n-1}/(n-1)!, & t > 0, \\ 0, & t \leq 0. \end{cases}$$

By the Young inequality, we have

$$\|k * u\|_p \leq \|k\|_1 \|u\|_p,$$

where $\|\cdot\|$ is a norm in $L_p(0,T)$. Since we have $\|k\|_1 = T^n/n!$, the rightmost estimate of (1.34) is established. The leftmost estimate is obtained after the substitution of the function $u(t) \equiv 1$ into the inequality $\|J^n u\|_p/\|u\|_p \leq \|J^n\|$. Using the fact that

$$(n!)^{1/n} \sim n/e, \qquad n^{1/n} \sim 1 \qquad n \to \infty$$

from the two-sided estimate (1.34) and (1.16), (1.17), we obtain

$$\rho(J) = 1, \qquad \sigma(J) = T. \qquad (1.36)$$

To generalize formulas (1.36) to the case of the Volterra operator (1.32) it is essential to use statement (ii) of Lemma 1.3, which implies that similar operators, i.e., the operators $A' = F^{-1}AF$, $F \in \mathcal{L}(X', X)$, $F^{-1} \in \mathcal{L}(X, X')$ have the same order and type. In this connection, the question arises what conditions we should impose on the kernel $A(t, \tau)$ defined in the triangle $\Delta = \{t, \tau \mid 0 < \tau < t < T\}$ in order that A be similar to J. The answer to this question is given by the following theorem.

Theorem 1.3. *Suppose that $A(t,\tau) \in C^2(\overline{\Delta})$, where $A(t,t) > 0$ for all $t \in [0,T]$. Then the operator A is similar to the operator cJ, where*

$$c = \frac{1}{T}\int_0^T A(t,t)\,dt.$$

From this theorem, taking into account Lemma 1.3 and formulas (1.36), we obtain the following corollary.

Corollary 1.1. *Under the hypotheses of Theorem 1.3, we have*

$$\rho(A) = 1, \qquad \sigma(A) = \int_0^T A(t,t)\,dt.$$

The algebraic scheme of the proof of Theorem 1.3 is based on the following lemma.

Lemma 1.4. *Suppose we have Banach spaces X and Y and a linear operator $J \in \mathcal{L}(X)$. We assume that there exist linear continuous mappings $\Gamma, \Phi \in \mathcal{L}(Y, \mathcal{L}(X))$ and a continuous bilinear mapping $\psi : Y \times Y \to Y$ such that for any $a, b \in Y$ we have*

$$[J, \Gamma(a)] = \Phi(a), \tag{1.37}$$

$$\Phi(\psi(a,b)) = \Phi(a)\Gamma(b), \tag{1.38}$$

and the operators $\Gamma(a)$ and $\Psi : b \to \psi(a,b)$ are quasinilpotent for all $a \in Y$. Then the operator $J + \Phi(a)$ is similar to J.

Proof. It suffices to construct the operator $F \in \mathcal{L}(X)$ such that

$$FJ = (J + \Phi(a))F \tag{1.39}$$

and to establish the existence of the bounded inverse operator $F^{-1} \in \mathcal{L}(X)$. We shall seek the operator F in the form

$$F = E + \Gamma(b), \tag{1.40}$$

where E is the unit operator. Substituting (1.40) into (1.39), we obtain

$$\Gamma(b)J = J\Gamma(b) + \Phi(a) + \Phi(a)\Gamma(b),$$

or, taking into account conditions (1.37) and (1.38),

$$\Phi(a + b + \psi(a,b)) = 0.$$

Chapter 4. Volterra operator equations and their applications

Thus, it suffices to find the solution $b \in Y$ of the equation

$$b + \psi(a, b) = -a.$$

The existence and uniqueness of a solution to this equation follows from the quasinilpotent property of the operator $\psi(a, \cdot)$. Substituting the element obtained above into $\Gamma(b)$, we find the operator F. The existence of the operator F^{-1} follows from the quasinilpotent property of the operator $\Gamma(b)$. The lemma is proved. □

Proof of Theorem 1.3. Considering the operator $c^{-1}A$ instead of A, we may assume that $c = 1$. First, we prove the theorem in the particular case

$$T = 1, \qquad A(t, t) = 1, \qquad (\partial A/\partial t)(t, t) = 0. \tag{1.41}$$

For this purpose, we write the operator A in the form

$$(Au)(t) = \int_0^t u(\tau)\,d\tau + \int_0^t a(t, \tau)u(\tau)\,d\tau = \mathcal{J}u + \Phi(a)u$$

where $a(t, \tau) = A(t, \tau) - 1$ and the operator $\Phi(a)$ is defined by the formula

$$\Phi(a)u = \int_0^t a(t, \tau)\,u(\tau)\,d\tau. \tag{1.42}$$

Moreover, taking into account (1.41), we have

$$a(t, t) = 0, \qquad (\partial_t a)(t, t) = 0. \tag{1.43}$$

We set $X = L_p(0, 1)$ in Lemma 1.4 and for the Banach space Y we take the class of functions $a(t, \tau)$ continuous in $\overline{\Delta}$, having continuous first and second derivatives with respect to t and satisfying conditions (1.43). We define the norm in the space Y by the formula

$$\|a\| = \sup |\partial_t^k a(t, \tau)|, \qquad (t, \tau) \in \overline{\Delta}, \quad k = 0, 1, 2.$$

For $a, b \in Y$ we set

$$(\Gamma(a)u)(t) = \int_0^t \left\{ \partial_t \partial_\tau \int_0^\tau a(s + t - \tau, s)\,ds \right\} u(\tau)\,d\tau, \tag{1.44}$$

$$(\psi(a, b))(t, \tau) = \int_\tau^t a(t, \eta) \left\{ \partial_\eta \partial_\tau \int_0^\tau b(s + \eta - \tau, s)\,ds \right\} d\eta. \tag{1.45}$$

The continuity of the operators $\Phi(a)$, $\Gamma(a)$, and the bilinear mapping $\psi(a,b)$ follows immediately from (1.42), (1.44), (1.45), and the definition of the norm in Y. From the definition of $\Gamma(a)$ we have

$$|(\Gamma(a)u)(t)| \leq \|a\| \int_0^t |u(\tau)|\,d\tau.$$

Hence, using formulas (1.33) and (1.35), by induction we obtain

$$|(\Gamma^n(a)u)(t)| \leq \|a\|^n (\mathcal{J}^n|u|)(t).$$

Therefore,
$$\|\Gamma^n(a)\| \leq \|a\|^n \|\mathcal{J}^n\|.$$

Thus, the operator $\Gamma(a)$ is quasinilpotent because so is the operator $\mathcal{J}$. We now prove that the operator $\Psi : b \to \psi(a,b)$ is quasinilpotent in Y.

Evidently,
$$|(\Psi b)(t,\tau)| \leq \|a\|\,\|b\|(t-\tau).$$

To show that ψ is quasinilpotent in Y, it suffices to prove the following estimate by induction:

$$|(\partial_t^k \Psi^n b)(t,\tau)| \leq \frac{(2\|a\|(t-\tau))^n}{n!}\|b\|, \quad k=0,1,2, \quad (t,\tau) \in \Delta. \quad (1.46)$$

To this end, differentiating the function in braces in formula (1.45) and applying the operator ∂_t^k to both sides of equality (1.45), in view of (1.43), we obtain

$$(\partial_t^k \psi)(t,\tau) = \int_\tau^t \partial_t^k a(t,\eta) \left\{ \partial_\eta b(\eta,\tau) - \int_0^\tau \partial_\eta^2 b(s+\eta-\tau,s)\,ds \right\} d\eta. \quad (1.47)$$

Hence, for $k=0,1,2$ we have

$$|(\partial_t^k \psi)(t,\tau)| \leq \|a\|\,\|b\| \int_\tau^t (1+\tau)\,ds \leq 2\|a\|\,\|b\|(t-\tau),$$

because $\tau \leq 1$, and thus estimate (1.46) is established for $n=1$. Assuming that this estimate holds for $k=n$ and substituting the function $\Psi^n b$ into formula (1.47) for the function $b(t,\tau)$, we have

$$|(\partial_t^k \Psi^{n+1} b)(t,\tau)| \leq \|a\| \int_\tau^t (1+\tau) \frac{(2\|a\|(t-\tau))^n}{n!}\|b\| \leq \frac{(2\|a\|(t-\tau))^{n+1}}{(n+1)!}\|b\|.$$

Chapter 4. Volterra operator equations and their applications

Thus, estimate (1.46) is established. Now, we show that identities (1.37) and (1.38) also hold. Taking the operator ∂_t in formula (1.44) out from the integral sign and carrying the operator ∂_τ over to the function u (we can perform these operations due to (1.43)), we obtain

$$(\Gamma(a)u)(t) = -\partial_t \int_0^t \left\{ \int_0^\tau a(s+t-\tau,s)\,ds \right\} \partial_\tau u(\tau)\,d\tau.$$

Hence,

$$(\mathcal{J}\Gamma(a)u)(t) = -\int_0^t \left\{ \int_0^\tau a(s+t-\tau,s)\,ds \right\} \partial_\tau u(\tau)\,d\tau$$

$$= \int_0^t a(t,\tau)u(\tau)\,d\tau + \int_0^t \left\{ \int_0^\tau \partial_\tau a(s+t-\tau,s)\,ds \right\} u(\tau)\,d\tau$$

$$= \int_0^t a(t,\tau)u(\tau)\,d\tau - \partial_t \int_0^t \int_0^\tau a(s+t-\tau,s)\,ds u(\tau)\,d\tau$$

$$= \Phi(a)u + (\Gamma(a)\mathcal{J}u)(t),$$

because $\partial_t a(s+t-\tau) = -\partial_\tau a(s+t-\tau)$.

Thus, $[\mathcal{J}, \Gamma(a)] = \Phi(a)$. By the definition of Φ and Γ, we have

$$(\Phi(a)\Gamma(b)u)(t) = \int_0^t a(t,\eta) \left\{ \int_0^\eta \left[\partial_\eta \partial_\tau \int_0^\tau b(s+\eta-\tau,s)\,ds \right] u(\eta)\,d\eta \right\} d\tau.$$

Changing the order of integration with respect to r and η, we obtain the identity

$$\Phi(a)\Gamma(b) = \Phi(\Psi(a,b))$$

since the function u is arbitrary. Thus, to finish the proof, we must show that the operator A from Theorem 1.3 for $c = 1$ is similar to an operator that satisfies condition (1.41) in the general case. For this purpose, we introduce an operator

$$F: L_p(0,1) \to L_p(0,T)$$

$$(Fv)(t) = e^{\alpha(t)} v(f(t)), \qquad t \in [0,T]$$

where

$$\alpha(t) = \int_0^t \frac{(\partial_t A)(\tau,\tau)}{A(\tau,\tau)}\,d\tau, \qquad f(t) = \int_0^t A(\tau,\tau)\,d\tau.$$

Since $f(0) = 0$, $f(T) = c = 1$, $f'(t) > 0$, the operator F has the inverse operator

$$(F^{-1}g)(t') = \exp(-\alpha(f^{-1}(t')))g(f^{-1}(t')).$$

It is easy to verify that the operator $A' = F^{-1}AF$ has the kernel

$$A'(t', \tau') = A(f^{-1}(t'), f^{-1}(\tau')) \exp(\alpha(f^{-1}(t')) - \alpha(f^{-1}(t'))) (f^{-1})'(\tau'),$$

$$0 < \tau' < t' \le 1$$

which satisfies condition (1.41). Theorem 1.3 is proved. □

The next theorem is the special version of Theorems 1.1, 1.2 in the case where $\Omega = (0, T)$, $\alpha = 1$, and the operator A is the Volterra operator (1.32) with real, continuous in $\overline{\Delta}$ and nonnegative kernel $A(t, \tau)$. Then condition (1.14) assumes the form

$$\|Vu(t)\|_{s'} \le (s - s')^{-1} \int_0^t A(t, \tau) \|u(\tau)\|_s \, d\tau. \tag{1.48}$$

Theorem 1.4. *Suppose $V \in \mathcal{L}(L_p(0, T; \mathcal{X}))$ and condition (1.48) holds for any s, $s \in I$, $s' < s$. Then (i) the solution of the equation*

$$u = Vu + f \tag{1.49}$$

is unique in $L_p(0, T; X_s)$, $s \in I$.

(ii) If the function f belongs to $L_p(0, T; X_s)$ for some $s, s' \in I$ and

$$e \int_0^T A(t, t) \, dt < s - s', \tag{1.50}$$

then equation (1.49) has a solution $u \in L_p(0, T; X_{s'})$ and $\|u\|_{p,s'} \le c\|f\|_{p,s}$, where c is independent of u and f.

(iii) If $\mathcal{X}$ is a Hilbert scale from Theorem 1.2 ($\alpha = 1$), $f \in L_p(0, T; X_0)$ and $\|\Lambda u\|_{p,0} \le m$, then

$$s > e \int_0^T A(t, t) \, dt, \qquad s \in I = (-a, a)$$

implies

$$\|u\|_{p,0} \le \omega_m(\|f\|_{p,0}),$$

where $\omega_s(\varepsilon) \sim sme/\ln\varepsilon^{-1}$ as $\varepsilon \to \infty$.

Proof. Since the function $A(t, \tau)$ is continuous and nonnegative, for any $\delta > 0$ there exists a function $A_\delta(t, \tau) \in C^2(\overline{\Delta})$ such that

$$\delta > A_\delta(t, \tau) - A(t, \tau) \ge 0, \quad 0 \le \tau \le t \le T, \qquad A_\delta(t, t) > 0, \qquad t \in [0, T].$$

Evidently, $V \in \mathcal{V}_p(\mathcal{X}, A_\delta, 1)$ and

$$\mathrm{e} \int_0^T A_\delta(t,t)\,\mathrm{d}t < s - s'$$

for sufficiently small δ. By Corollary 1.1, we have

$$\rho(As) = 1, \qquad \sigma(A_\delta) = \int_0^T A_\delta(t,t)\,\mathrm{d}t,$$

where A_δ is a Volterra operator in $L_p(0,T)$ with kernel $A_\delta(t,\tau)$. Using Theorems 1.1 and 1.2 and passing to the limit as $\delta \to 0$, we obtain statements (ii) and (iii).

Statement (i) is proved under the additional condition (1.50) which holds a priori for small T. Since we can move along the t-axis by small steps when proving the uniqueness, repeating this argument finite number of times we prove uniqueness in the whole. The theorem is proved. □

Exercise 1.1. In the proof of Lemma 1.1 we only show that $u = 0$ in the space $X_{s'}$. Why $u = 0$ in the space X_s?

Exercise 1.2. Deduce the estimate for $r(s, s', V)$ which was used in the proof of Theorem 1.1.

Exercise 1.3. In Example 1.2, suppose that $\lambda(\xi) = |\xi|^{1/2}$, $\alpha \geq 1$. Prove that

$$\|\partial_j u\|_{s'} \leq \alpha(\mathrm{e}(s-s'))^{-\alpha}\|u\|_s, \qquad s' < s.$$

Exercise 1.4. Suppose Ω is an open set in $\mathbb{R}^n$. We denote by X_s the closure of the subset of functions from $C^\infty(\Omega)$ for which the norm

$$\|u\|_s = \sup_{n \geq 0}((s/n)^{an}|u|_n), \qquad \alpha > 0, \quad s > 0,$$

is finite, where

$$|u|_n = \sup_{\substack{x \in \Omega \\ |\rho|=n}} |\partial^\rho u(x)|$$

(for $n = 0$ we set $(s/n)^{an} = 1$). Prove that the set X_s, $s \in (0,r)$ forms a scale of Banach spaces and for $s' < s$

$$\|\partial_j u\|_{s'} \leq c(r)(s-s')^{-\alpha}\|u\|_s.$$

What conditions should the function $a \in C^\infty(\Omega)$ satisfy in order for the operator of multiplication by this function to be bounded in X_s? Consider the cases $\alpha < 1$, $\alpha = 1$, and $\alpha > 1$.

Exercise 1.5. Let H and W be separable Hilbert spaces such that W is densely embedded in H and $\|u\|_W = \|\Lambda u\|$, where Λ is a self-adjoint positive operator in H, $\Lambda : W \to H_j$, and $\|\cdot\|$ is a norm in H. Consider a two-parameter set of operators

$$V(t,\tau) : W \to H, \qquad (t,\tau) \in \Omega = \{t, \tau \in \mathbb{R} \mid 0 < \tau < t < T\}$$

such that for all $u \in W$, $(t,\tau) \in \Omega$, $s \in (-m, m)$ the following estimate holds:

$$\|\exp(s\Lambda^\nu) V(t,\tau) \exp(-s\Lambda^\nu) u\| \le c(t-\tau)^{\alpha-1} \|\Lambda u\|.$$

Here, the positive constants α, ν, c, m are independent of t, τ, u. Prove the following statements for the equation

$$u(t) + \int_0^t V(t,\tau) u(\tau)\, d\tau = f(t). \tag{1.51}$$

(i) The solution of equation (1.51) is unique in the space $L_p(0, T; W)$ for $\alpha\nu \ge 1$.

(ii) If $\alpha\nu > 1$ or $\alpha\nu = 1$ and $m > T(e/\alpha)^{1-1/\alpha}(c\Gamma(\alpha))^{1/\alpha}$, then equation (1.51) is densely solvable, i.e., it has a solution $u \in L_p(0, T; W)$ for a set of right-hand sides f which is dense in $L_p(0, T; H)$.

(iii) For any $\varepsilon > 0$, the following estimate of (conditional) stability holds:

$$\|u\|_{L_p(0,T;H)} \le \varepsilon \|u\|_{L_p(0,T;W)} + c_1 \exp(s\varepsilon^{-\nu}) \|f\|_{L_p(0,T;H)},$$

where the number c_1 is independent of ε and u and the number s satisfies the condition $0 < s < m$ for $\alpha\nu > 1$ and $m > s > T(e/\alpha)^{1-1/\alpha}(c\Gamma(\alpha))^{1/\alpha}$ for $\alpha\nu = 1$. Here $\Gamma(\alpha)$ is the gamma function.

4.2. NONHYPERBOLIC CAUCHY PROBLEM FOR THE WAVE EQUATION

We shall obtain the estimates of stability for the problem of extending a solution of the hyperbolic equation

$$u_{tt} - \Delta u + a(x,t) u = f(x,t), \qquad x \in \mathbb{R}^2, \quad |t| < T, \tag{2.1}$$

Chapter 4. Volterra operator equations and their applications 111

to the exterior of the time-like cylinder $\Gamma = \{(x,t) \mid |x| = 1, \ |t| < T|\}$ with zero Cauchy data
$$u|_\Gamma = \partial u/\partial \nu|_\Gamma = 0. \tag{2.2}$$

Here ν is the unit normal to Γ. In order to formulate the main result, it is convenient to rewrite this problem in the polar coordinate system $x_1 = r\cos\varphi$, $x_2 = r\sin\varphi$:

$$u_{tt} - r^{-1}\partial_r(r\partial_r u) - r^{-2}\partial_\varphi^2 u + au = f, \tag{2.3}$$

$$u|_{r=1} = \partial_r u|_{r=1} = 0. \tag{2.4}$$

Here the functions $u = u(r,t,\varphi)$, $a = a(r,t,\varphi)$ and f are 2π-periodic with respect to φ. Set

$$\Omega = \{r, t \mid r \in [1, T+1], \quad |t| < T + 1 - r\}.$$

Theorem 2.1. *Suppose the function $a(r,t,\varphi)$ is analytic with respect to φ in the strip $|\operatorname{Im}\varphi| < s$ and bounded in its closure uniformly with respect to $(r,t) \in \pi$, where*

$$s > \ln(1+T). \tag{2.5}$$

If $\|u_{\varphi\varphi}\| \leq m$ in the norm $L_2(\Omega \times (0, 2\pi))$, then a solution of problem (2.3), (2.4) is unique and the following stability estimate holds:

$$\|u\| \leq \omega_m(\|f\|), \tag{2.6}$$

$$\omega_m(\delta) \sim ms^2/\ln^2 \delta^{-1}, \quad \delta \to 0.$$

The proof of this estimate is based on Theorem 1.2 with the corresponding operator A. The basic difficulty consists in derivation of the number $\sigma(A)$. Note that the estimate of the number $\sigma(A)$ based on the estimate from above of the kernel of the operator A leads to the condition of the form $s > \exp(cT)$, while the analysis carried out below allows replacing it by (2.5).

Now we make auxiliary constructions and estimates. Set

$$L_a u = u_{tt} - r^{-1}\partial_r(r\partial_r u) + a(r,t,\varphi)u. \tag{2.7}$$

The hyperbolic operator L_a for smooth functions in $\overline{\Omega}$ which satisfy initial data (2.4) has inverse operator bounded in the space L_2. Therefore, problem (2.3), (2.4) is equivalent to solution of the integro-differential equation

$$u = L_a^{-1} r^{-2} \partial_\varphi^2 u + L_a^{-1} f. \tag{2.8}$$

So, in the following, we define a solution of problem (2.3), (2.4) as a solution of equation (2.8), and the function u is not necessarily assumed to be differentiable with respect to r and t.

To apply Theorem 1.2 to estimate (2.8), we introduce the corresponding scale of Hilbert spaces.

In Theorem 1.2, let $X \equiv X_0$ be the space of 2π-periodic functions $u(\varphi) \in L_2(0, 2\pi)$; $\Lambda = -\partial_\varphi^2$, $\alpha = 2$. Using the expansion into the Fourier series and the Parseval equality, we obtain

$$\|u\|_s^2 = \sum_{n=-\infty}^{\infty} e^{2s|n|} |\hat{u}_n|^2, \qquad \|\cdot\|_s \equiv \|\cdot\|_{X_s},$$

$$\hat{u}_n = \frac{1}{2\pi} \int_0^{2\pi} u(\varphi) \exp(-in\varphi) \, d\varphi.$$

As in Example 1.2, for the 2π-periodic function $a(\varphi + i\psi)$ analytic in the strip $|\psi| < s$ and bounded in its closure, we have

$$\|au\|_s \leq \bar{a} \|u\|_s, \tag{2.9}$$

$$\bar{a} = \sqrt{2} \sup |a(\varphi + i\psi)|, \qquad \varphi \in [0, 2\pi], \quad |\psi| < s$$

and, moreover,

$$\|\partial^2 \varphi u\|_{s'} \leq 4e^{-2}(s - s')^{-2} \|u\|_s. \tag{2.10}$$

Setting

$$V = L_a^{-1} r^{-2} \partial_\varphi^2, \qquad g = L_a^{-1} f \tag{2.11}$$

we rewrite the problem (2.8) in the abstract form $u = Vu + g$. To estimate $\|Vu\|_{s'}$, we find out the structure of the operator L_a^{-1}. Denote the variable t by x and introduce the new variable $t = r - 1$ instead of r. Let $u = \mathcal{F}v$, $\mathcal{F} = F(1+t)^{-1/2}$, where F is the operator of transition from the variable t to $r = t + 1$. Then

$$L_a \mathcal{F} v = -\mathcal{F} \square_b v,$$

where

$$\square_b v = v_{tt} - v_{xx} - b(t+1, x, \varphi) v(x, t),$$

$$b(r, x, \varphi) = a(r, x, \varphi) - (2r)^{-2}.$$

In other words, $L_a = -\mathcal{F} \square_b \mathcal{F}^{-1}$,

$$L_a^{-1} = -\mathcal{F} \square_b^{-1} \mathcal{F}^{-1}. \tag{2.12}$$

Chapter 4. Volterra operator equations and their applications

Since $\Box_b = \Box - b$, we have

$$\Box_b^{-1} = \sum_{n=0}^{\infty} (b\Box^{-1})^n \Box^{-1}, \tag{2.13}$$

$$(\Box^{-1} u)(x,t) = \frac{1}{2} \int_\Delta \Theta(t - \tau - |x - \xi|) u(\xi, \tau) \, d\xi \, d\tau,$$

$$\Delta = \{x, t \mid t \in [0, T], \ |x| < T - t\}.$$

As the operator $\Box^{-1}$ is a Volterra operator, the series (2.13) represents the Volterra operator in each of the spaces $L_p(\Delta)$ and $C(\overline{\Delta})$ if b is bounded. (Hereinafter we use the term "Volterra operator" as a synonym of the term "quasinilpotent operator".) Evidently, for $u \in L_2(\Delta; X_s)$

$$\|\Box^{-1} u\|_s \leq \Box^{-1} \|u\|_s. \tag{2.14}$$

Combining estimates (2.14) and (2.9) and using induction, we obtain

$$\|(b\Box^{-1})^n u\|_s \leq (\bar{b}\Box^{-1})^n \|u\|_s, \tag{2.15}$$

$$\bar{b} = \sqrt{2} \sup |b(t+1, x, \varphi + i\psi)|, \qquad (x,t) \in \Delta,$$

$$\varphi \in [0, 2\pi), \qquad |\psi| < s_0, \qquad s \leq s_0.$$

Formulas (2.13)–(2.15) yield

$$\|\Box_b^{-1} u\|_s \leq \Box_c^{-1} \|u\|_s, \qquad c = \bar{b}. \tag{2.16}$$

Since $\|\mathcal{F}^{\pm 1} u\|_s = \mathcal{F}^{\pm 1} \|u\|_s$, from (2.10)–(2.12) and (2.16) we have the estimate

$$\|Vu\|_s = \|\mathcal{F} \Box_b^{-1} (1+t)^{-2} \mathcal{F}^{-1} \partial_\varphi^2 u\|_{s'} \leq (s - s')^{-2} A \|u\|_s,$$

where A is a Volterra operator in $L_2(\Omega)$:

$$A = 4e^{-2} \mathcal{F} \Box_c^{-1} (1+t)^{-2} \mathcal{F}^{-1}.$$

Therefore, by Lemma 1.3, we have

$$\rho(A) = \rho(B), \qquad \sigma(A) = (4e^{-2})^{\rho(B)} \sigma(B), \tag{2.17}$$

where $B = \Box_c^{-1} (1+t)^{-2}$ is a Volterra operator in $L_2(\Delta)$.

Further, we shall need the following statement.

Lemma 2.1. *Suppose X_1, X_2 are Banach spaces and an operator A belongs to $\mathcal{L}(X_1) \cap \mathcal{L}(X_2)$. If $A^m \in \mathcal{L}(X_1, X_2)$, $A^p \in \mathcal{L}(X_2, X_1)$, and $\mathrm{Sp}_1 A = \{0\}$ for some integers $m, p \geq 1$, ($\mathrm{Sp}_j A$ is the spectrum of the operator A in X_j), then $\mathrm{Sp}_2 A = \{0\}$ and the numbers $\rho(A)$ and $\sigma(A)$ are independent of whether the operator A is considered in X_1 or in X_2.*

Proof. Set
$$c = \max\{\|A^m\|_{\mathcal{L}(X_1, X_2)}, \|A^p\|_{\mathcal{L}(X_2, X_1)}\},$$

$$f_j(\lambda) = \sum_{n=0}^{\infty} \|A^n\|_{\mathcal{L}(X_j)} \lambda^n, \qquad p_j(\lambda) = \sum_{n=0}^{m+p-1} \|A^n\|_{\mathcal{L}(X_j)} \lambda^n, \qquad j = 1, 2.$$

For $n \geq m + p$, from the formula $A^n = A^p A^{n-m-p} A^m$ it follows that

$$\|A^n\|_{\mathcal{L}(X_1)} \leq c^2 \|A^{n-m-p}\|_{\mathcal{L}(X_2)}, \tag{2.18}$$

$$f_1(\lambda) \leq p_1(\lambda) + c^2 \lambda^{m+p} f_2(\lambda), \qquad \lambda \geq 0.$$

If we interchange X_1 and X_2, we obtain

$$f_2(\lambda) \leq p_2(\lambda) + c^2 \lambda^{m+p} f_1(\lambda), \qquad \lambda \geq 0. \tag{2.19}$$

So, since the order and the type of an entire function do not change when multiplying by or summing with a polynomial, the functions f_1 and f_2 have the same order and type, which is required.

Using the well-known formula for $\square_c^{-1}$, we obtain

$$(Bu)(x,t) = \frac{1}{2} \int_0^t \int_{x-t+\tau}^{x+t-\tau} J_0\sqrt{ic[(t-\tau)^2 + (x-\xi)^2]}(1+\tau)^{-2} u(\xi, \tau) \, d\xi \, d\tau,$$

where J_0 is the Bessel function. Hence, we see that for the operator B the assumptions of Lemma 2.1 hold with $X_1 = L_2(\Delta)$, $X_2 = C(\overline{\Delta})$, $m = p = 1$. Therefore, it suffices to determine the order and type of the operator B in $c(\overline{\Delta})$. We show that $\rho(B) = \rho(B_1)$ and $\sigma(B) = \sigma(B_1)$, where B_1 is the restriction of the operator B to the invariant subspace consisting of the functions independent of x, i.e., to $C[0,T]$. Indeed, it is evident that $\|B_1^n\| \leq \|B^n\|$. On the other hand, since the kernel of B is nonnegative, we have

$$|(Bu)(x,t)| \leq (B\bar{u})(t) = (B_1\bar{u})(t),$$

where $\bar{u}(t) = \sup |u(x,t)|$. Hence, $\|B^n u\| \leq \|B_1^n \bar{u}\|$ or $\|B^n\| \leq \|B_1^n\|$. Since $B_1 = (\partial_t^2 - c)^{-1}(t+1)^{-2}$, we obtain

$$B_1^{-1} = (t+1)^2 (\partial_t^2 - c),$$

Chapter 4. Volterra operator equations and their applications   115

$$\mathcal{D}(B_1^{-1}) = \{u \mid u \in C^2[0,T], \ u(0) = u'(0) = 0\} \equiv \mathcal{D}.$$

It is easy to see that the operator L of the form

$$Lu = a^2(t)u_{tt} + b(t)u_t + c(t)u$$

with domain of definition $\mathcal{D}$ is similar to the operator $\lambda^2 \partial_t^2$,

$$\lambda = \frac{T}{\int_0^T ds/a(s)},$$

if $a(t) \geq \delta > 0$ and the functions a, b, and c are sufficiently smooth. To prove this fact, it suffices to make the standard change of variables

$$u(t) = \Phi \delta = e^{h(t)} v(\varphi(t)),$$

where $\varphi = \lambda \int_0^t ds/a(s)$ and the function h is a solution to the first order equation

$$a^2[2h'\varphi' + \varphi''] + b\varphi' = 0.$$

Then $L\Phi v = \lambda^2 \Phi L_0 v$, i.e., $L = \lambda^2 \Phi L_0 \Phi^{-1}$, where

$$L_0 = \partial_t^2 + q(t), \qquad q = \lambda^{-2}(h'^2 + h'' + c) \circ \varphi^{-1},$$

and the similarity of L_0 to the operator ∂_t^2 follows from Corollary 3.1, Chapter 2. If $L = B_1^{-1}$, then the number λ is equal to $\ln^{-1}(1+T)$. Consequently, the operator B_1 is similar to $\lambda^{-2} J^2$. Therefore, Lemma 1.3, formula (2.17), and the equalities $\sigma(J) = T$ and $\rho(J) = 1$ yield

$$\rho(A) = 1/2, \qquad \sigma(A) = 2e^{-1} \ln(1+T).$$

Thus, Theorem 1.2 implies Theorem 2.1.   □

4.3. THE PROBLEM OF INTEGRAL GEOMETRY IN A STRIP

Suppose that in the strip $\Omega = \{\xi, \eta \mid \xi \in \mathbb{R}, \ \eta \in (0, h)\}$ we have the set of parabolas

$$\gamma(x,y) : \eta = y - (\xi - x)^2, \qquad \xi \in (\xi_1, \xi_2)$$

with vertices $(x, y) \in \Omega$ and ends lying in the axis $\eta = 0$ (see Figure 7).

For a continuous weight function $a(\xi, x, y)$ and a finite continuous function $u \in C_0(\Omega)$, we set

$$(Pu)(x,y) = \int a(\xi, x, y) u(\xi, y - (\xi - x^2)) \, d\xi = g(x, y). \qquad (3.1)$$

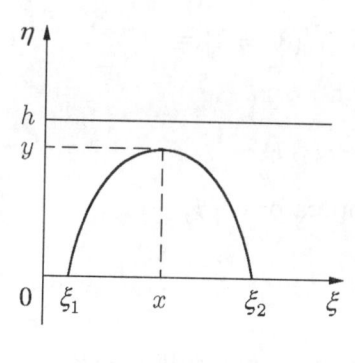

Figure 7

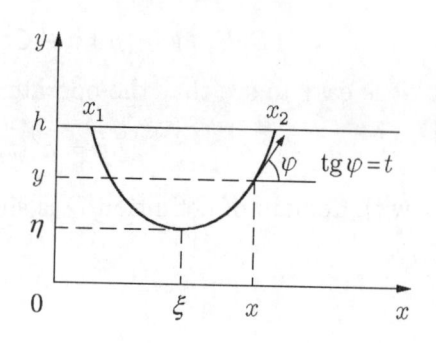

Figure 8

Given a function g defined in Ω, it is required to reconstruct the function u. The problems in which we need to find the function $u(x,y)$ knowing its integrals with a certain weight along a certain set of curves are called the problems of integral geometry. In particular, if

$$a(\xi, x, y) = (1 + 4(\xi - x)^2)^{1/2},$$

then equation (3.1) assumes the form

$$\int_{\gamma(x,y)} u\, ds = g(x, y),$$

where ds is the Euclidean element of length of the curve γ. Our goal is to prove the uniqueness of the solution of the equation $Pu = g$. The proof is based on the existence of a solution of the conjugate problem, which, in its turn, follows from Theorem 1.1. First, we show that the operator P is extended by continuity to the space $L_1(\Omega)$ and find the conjugate operator.

Lemma 3.1. *Suppose that*

$$|a(\xi, x, y)| \leq m \qquad (3.2)$$

for all $\xi, x \in \mathbb{R}$, $y \in (0, h)$. Then

$$\|Pu\|_1 \leq 2\sqrt{hm}\|u\|_1, \qquad (3.3)$$

for all $u \in L_1(\Omega)$, where $\|\cdot\|_1$ is the norm in $L_1(\Omega)$.

Proof. In order not to care about the limits of integration, we write the operator P in the form

$$(Pu)(x, y) = \int \delta(\eta - y + (\xi - x)^2) a(\xi, x, y) u(\xi, \eta)\, d\xi\, d\eta, \qquad (3.4)$$

Chapter 4. Volterra operator equations and their applications

where δ is the Dirac function concentrated on the curve γ, $u \in C_0(\Omega)$. Taking into account (3.2), we obtain

$$|(Pu)(x,y)| \leq m \int \delta(\eta - y + (\xi - x)^2) u(\xi, \eta) \, d\xi \, d\eta.$$

Changing the order of integration, from this inequality we obtain

$$\|(Pu)(x,y)\|_1 \leq m \int |u(\xi, \eta)| \left\{ \int_\Omega \delta(\eta - y + (\xi - x)^2) \, dx \, dy \right\} d\xi \, d\eta$$
$$\leq 2m\sqrt{h} \|u\|_1,$$

since the integral in braces is equal to

$$\int_{x_1}^{x_2} dx = x_2 - x_1 \leq 2\sqrt{h}.$$

Here x_1, x_2 are the points of intersection of the parabola $\eta - y + (\xi - x)^2 = 0$ with the line $y = h$ in the plane x, y (Figure 8). □

Since the space $C_0(\Omega)$ is dense in $L_1(\Omega)$, estimate (3.3) is proved. We shall need the set of parabolas

$$\gamma^*(\xi, \eta) : y = \eta + (\xi - x)^2, \quad x \in (x_1, x_2),$$

which arose in the proof, to find P^*. For $v \in C_0(\Omega)$, we set

$$(P^*v)(\xi, \eta) = \int b(x, \xi, \eta) v(x, \eta + (\xi - x)^2) \, dx, \qquad (3.5)$$

where

$$b(x, \xi, \eta) = a(\xi, x, \eta + (\xi - x)^2). \qquad (3.6)$$

We show that for all $u, v \in C_0(\Omega)$

$$\langle Pu, v \rangle = \langle u, P^*v \rangle, \qquad (3.7)$$

where $\langle \cdot, \cdot \rangle$ is the scalar product in $L_2(\Omega)$. Indeed, using formula (3.4), we have

$$\langle Pu, v \rangle = \int \left\{ \int \delta(\eta - y + (\xi - x)^2) a(\xi, x, y) u(\xi, \eta) \, d\xi \, d\eta \right\} v(x, y) \, dx \, dy$$
$$= \int u(\xi, \eta) \left\{ \int \delta(\eta - y + (\xi - x)^2) a(\xi, x, y) v(x, y) \, dx \, dy \right\} d\xi \, d\eta$$
$$= \langle u, P^*v \rangle.$$

Here we have changed the order of integration and avoided integration with respect to y in the interior integral. Under condition (3.2), formula (3.7) is extended to $u \in L_1(\Omega)$ and $v \in C(\Omega) \cap L_\infty(\Omega)$ by continuity.

Further, it is convenient to consider the curve $\gamma^*(\xi, \eta)$ as a solution of the Cauchy problem

$$y''(x) = 2, \quad y|_{x=\xi} = \eta, \quad y'|_{x=\xi} = 0. \tag{3.8}$$

We shall also need the following mappings connected with the curves $\gamma^*(\xi, \eta)$.

The mapping $\Phi : \Omega \to \mathbb{R}^2$,

$$\Phi : (\xi, \eta) \to (\eta + (x - \xi)^2, 2(x - \xi))(y, t),$$

which maps the vertex (ξ, η) of the parabola $\gamma^*(\xi, \eta)$ to the ordinate of the point $(x, y) \in \gamma^*(\xi, \eta)$ and the tangent $y'(x) = t$ of the slope of the curve γ^* at this point (see Figure 8). It depends on the parameter $x \in \mathbb{R}$. The inverse mapping, evidently, is given by the formula

$$(\xi, \eta) = \Phi^{-1}(y, t) = (x - t/2, y - t^2/4).$$

The mappings $\Phi_j : \Omega \to \Omega_j$ are given by the formulas

$$\Phi_j : (\xi, \eta) \to (x_j(\xi, \eta), t_j(\xi, \eta)),$$

where

$$x_j(\xi, \eta) = \xi + (-1)^j \sqrt{h - \eta}, \quad t_j(\xi, \eta) = (-1)^j 2\sqrt{h - \eta},$$

$$\Omega_1 = \mathbb{R} \times (-2\sqrt{h}, 0), \quad \Omega_2 = \mathbb{R} \times (0, 2\sqrt{h}).$$

The mapping Φ_j maps the vertex (ξ, η) of the curve $\gamma^*(\xi, \eta)$ to one of its ends and the tangent of the slope of the curve $\gamma^*(\xi, \eta)$ at this point. The inverse mapping is given by the formula

$$(\xi, \eta) = \Phi_j^{-1}(x_j, t_j) = (x_j - t_j/2, h - t_j^2/4).$$

We set $\rho(x, y, t) = b \circ \Phi^{-1} = b(x, x - t/2, y - t^2/4)$. In terms of the initial weight a, taking into account formula (3.6) and the fact that $(\xi - x)^2 = t^2/4$, we have

$$\rho(x, y, t) = a(x - t/2, x, y). \tag{3.9}$$

Suppose $w(x, y, t)$ is a solution of the Cauchy problem

$$Lw \equiv tw_y + w_x + 2w_t = \rho(x, y, t)v(x, y), \tag{3.10}$$

Chapter 4. Volterra operator equations and their applications 119

$$w|_{y=h} = f(x,t). \tag{3.11}$$

Note that, as it follows from (3.8), we have

$$(Lw)(x,y,y'(x)) = \frac{\mathrm{d}}{\mathrm{d}x}w(x,y,y'(x)) \tag{3.12}$$

along the curve $\gamma^*(\xi,\eta)$. Taking into account (3.5), (3.10)–(3.12), we obtain

$$(P^*v) = \int \rho(x,y(x),y'(x))v(x,y(x))\,\mathrm{d}x = \int_{x_1}^{x_2} \frac{\mathrm{d}}{\mathrm{d}x} w(x,y(x),y'(x))\,\mathrm{d}x$$

$$= f(x_2(\xi,\eta),t_2(\xi,\eta)) - f(x_1(\xi,\eta),t_1(\xi,\eta))$$

$$= f \circ \Phi_2 - f \circ \Phi_1 = Q_2 f - Q_1 f = Qf,$$

where $Q_j f(\xi,\eta) = f(\Phi_j(\xi,\eta))$ is the superposition operator and the operator $Q = Q_2 - Q_1$. Thus, formula (3.7) yields the identity

$$\langle Pu, v\rangle = \langle u, Qf\rangle,$$

where the function v is connected with f by conditions (3.10) and (3.11). We now find the conjugate operator Q^*. For $u \in C_0(\Omega)$ we have

$$\langle u, Qf\rangle = \sum_{j=1}^{2}(-1)^j \int_{\Omega} u(\xi,\eta)f(\Phi_j(\xi,\eta))\,\mathrm{d}\xi\,\mathrm{d}\eta$$

$$= \sum_{j=1}^{2}(-1)^j \frac{1}{2}\int_{\Omega_j} u(\Phi_j^{-1}(x_j,t_j))f(x_j,t_j)|t_j|\,\mathrm{d}x_j\,\mathrm{d}t_j,$$

since

$$\frac{\partial(\xi,\eta)}{\partial(x_j,t_j)} = \begin{vmatrix} 1 & 0 \\ -1/2 & -t_j/2 \end{vmatrix} = -\frac{t_j}{2}.$$

Let χ_j be the characteristic function of the set Ω_j. Then

$$\langle u, Qf\rangle_{L_2(\Omega)} = \int_{\Omega_1\cup\Omega_2} \frac{1}{2}\sum_{j=1}^{2}(-1)^j\chi_j(x,t)u(\Phi_j^{-1}(x,t))|t|f(x,t)\,\mathrm{d}x\,\mathrm{d}t$$

$$= \langle Q^*u, f\rangle_{L_2(\Omega_1\cup\Omega_2)},$$

where

$$Q^*u = \frac{1}{2}\sum_{j=1}^{2}(-1)^j\chi_j(x,t)u(\Phi_j^{-1}(x,t))|t|. \tag{3.13}$$

Making the inverse change $\Phi_j^{-1}(x,t) = (\xi, \eta)$, we find

$$\|Q^*u\|_{L_1(\Omega_1 \cup \Omega_2)} = \frac{1}{2} \sum_{j=1}^{2} \int_{\Omega_t} |u(\Phi_j^{-1}(x,t))| |t| \, dx \, dt$$

$$= \sum_{j=1}^{2} \|u\|_{L_1(\Omega)} = 2\|u\|_{L_1(\Omega)}.$$

Thus, we have proved that $Q^*/2$ is an isometric operator acting from $L_1(\Omega)$ into $L_1(\Omega \cup \Omega_2)$. The final point of our calculations is the following lemma.

Lemma 3.2. *For any $u \in L_1(\Omega)$ $f \in C(\overline{\Omega_1 \cup \Omega_2})$, the identity*

$$\langle Pu, v \rangle = \langle Q^*u, f \rangle \tag{3.14}$$

is valid under the condition that the function v of class $C(\overline{\Omega})$ is connected with f by relations (3.10) and (3.11), and the operator Q^* is defined by formula (3.13). In this case,

$$\|Q^*u\|_1 = 2\|u\|_1. \tag{3.15}$$

Now, we investigate the problem of determination of the functions $v(x,y)$, $w(x,y,t)$ from conditions (3.10), (3.11). Setting $t = 0$ in equation (3.10) under the condition that $\rho(x,y,0) \neq 0$, we may express the function v in terms of the function w:

$$v(x,y) = \rho^{-1}(x,y,0)(w_x + 2w_t)|_{t=0} = \rho^{-1} Lw|_{t=0}.$$

As a result, equation (3.10) can be written as follows:

$$tw_y = \rho^{-1}(w_x + 2w_t)|_{t=0}\rho(x,y,t) - (w_x + 2w_t).$$

Applying the difference operator

$$\Delta w(x,y,t) = (w(x,y,t) - w(x,y,0))/t$$

to both sides of this equation, we obtain

$$w_y = \rho^{-1}(w_x + 2w_t)|_{t=0}\Delta\rho - \Delta(w_x + 2w_t).$$

Hence, integrating with respect to y from y to h and taking into account conditions (3.9) and (3.11), we obtain

$$f(x,t) - w(x,y,t)$$
$$= \int_y^h \{a^{-1}(x,x,\eta)(w_x(x\eta,0) + 2w_t(x,\eta,0))\Delta a(x-t/2,x,\eta)$$
$$- \Delta(w_x + 2w_t)(x,\eta,t)\} \, d\eta \stackrel{\text{def}}{=} -Vw$$

Chapter 4. Volterra operator equations and their applications   121

or, in a brief form,
$$w = Vw + f.$$

We may consider the right-hand side $f(x,t)$ as given in the strip $\mathbb{R} \times (-2\sqrt{h}, 2\sqrt{h})$. For the Banach space X_s we take the space of functions $f(x,t)$ analytic in the complex domain

$$\mathcal{D} = \{x, t \in \mathbb{C} \mid |\operatorname{Im} x| \leq \varepsilon(1+s), \quad |t| < 2\sqrt{h} + \varepsilon(1+s)\}$$

and bounded in its closure,

$$\|f\|_s = \sup |f(x,t)|, \qquad (x,t) \in \mathcal{D}_s.$$

Here ε is a fixed positive number, $s \in [0,1]$. Suppose that for all $y \in [0,h]$ we have $a(x - t/w, x, y) \in X_1$ and

$$|a(x - t/2, x, y)| \leq \alpha, \qquad (x,y) \in \mathcal{D}_1, \tag{3.16}$$

and, moreover,

$$|a(x, x, y)| \geq \rho > 0, \qquad (x, 0) \in \mathcal{D}_1. \tag{3.17}$$

Following Example 1.1 and the definition of the operator V, we have

$$\|\partial w\|_{s'} \leq (\varepsilon(s - s'))^{-1}\|w\|_s, \qquad w \in X_s, \tag{3.18}$$

for $s' < s$, where ∂ is the operator of differentiation either with respect to x or with respect to t. To estimate $\|\Delta w\|_s$, $w \in X_s$, we note that the function Δw is analytic with respect to t in the disk $|t| < 2\sqrt{h} + \varepsilon(1+s)$ for fixed x and, consequently, it reaches its maximum on the boundary $|t| = 2\sqrt{h} + \varepsilon(1+\varepsilon) \geq 2\sqrt{h}$. Hence,

$$\|\Delta w\|_s \leq h^{-1/2}\|w\|_s. \tag{3.19}$$

Using estimates (3.16)–(3.19) and the definition of the operator V, for $s' < s$ we obtain

$$\|Vw\|_{s'} \leq c(s - s')^{-1} \int_y^h \|w(\cdot, \eta, \cdot)\|_s \, d\eta,$$

where $c = 3h^{-1/2}\varepsilon^{-1}(\alpha/\rho + 1)$.

Suppose that an operator A acts in the space $C[0, h]$ and is defined by the formula

$$(A\varphi)y = c \int_y^h \varphi(\eta) \, d\eta.$$

It is easy to see that $\rho(A) = 1$, $\sigma(A) = ch$.

Theorem 1.1 for the scale $C([0, h]; X_s)$ (in this case it is proved in the same way as for $L_\infty(0, h; X_s)$), with $f \in X_1$, asserts that there exists a solution $w(x, y, t) \in C([0, h]; X_0)$ if the number $ech < 1$. Moreover,

$$\|w\|_{C([0,h];X_0)} \leq c_1 \|f\|_{C([0,h];X_1)}.$$

Continuous differentiability of the function w with respect to y follows immediately from the equation $w = Vw + f$. Performing the calculations which lead us to this equation in the reverse direction and setting

$$v(x, y) = a^{-1}(x, x, y)(w_x + 2w_t)|_{t=0},$$

we obtain relations (3.10) and (3.11).

Now, using Lemma 3.2, it is easy to prove the following uniqueness theorem for the equation $Pu = g$.

Theorem 3.1. *Suppose that a function $a(\xi, x, y)$ is continuous with respect to $y \in [0, h]$ and analytic with respect to ξ, x in the complex domain*

$$|\operatorname{Im} x| < \delta, \qquad |\xi - x| < \delta, \qquad x, \xi \in \mathbb{C},$$

and there exists a number $\rho > 0$ such that $|a(x, x, y)| \geq \rho$ for all $x \in \mathbb{C}$, $|\operatorname{Im} x| < \delta$, $y \in [0, h]$. Then the solution of the equation $Pu = g$ is unique in $L_1(\Omega)$.

Proof. Diminishing the numbers h and ε if necessary, we may satisfy conditions (3.16), (3.17), and the inequality $ech < 1$. Then, as it was proved above, for $f \in X$ there exists a function $v \in C(\overline{\Omega})$ such that identity (3.14) holds. For $Pu = 0$ this identity takes the form $\langle Q^*u, f \rangle = 0$ for all $f \in X_1$. Since the set of analytic functions is dense in the space $C(\overline{\Omega})$, it follows that $Q^*u = 0$. Since we have $\|Q^*u\|_1 = 2\|u_1\|$ by Lemma 3.2,

$$u(x, y) = 0, \qquad x \in \mathbb{R}, \quad y \in [0, h].$$

Repeating the above considerations for the intervals $[h, 2h], [2h, 3h]$, and so on, we obtain the uniqueness theorem for any finite h. The theorem is proved. □

Exercise 3.1. Suppose $a(\xi, x, y) \equiv 1$. Prove that the solution of the equation $Pu = g$ is given by the formula

$$u(x, y) = \frac{1}{\pi} \int_0^y \frac{\cos(\partial_x \sqrt{y - s})}{\sqrt{y - s}} \partial_s g(x, s) \, ds.$$

Here $\cos(\partial_x \sqrt{y-s})$ is the differential operator of infinite order with respect to x. For what functions $g(x,y)$ does this formula guarantee the existence of a continuous solution $u(x,y)$?

Exercise 3.2. Using identities (3.14) and (3.15) and setting $f(x,t) = \exp(ix\alpha + it\mu)$, obtain the stability estimates for solutions of the equation $Pu = g$ if $u \in C_0(K)$, where K is a fixed compact set in Ω and $\|u\|_{C^1} \leq m$.

4.4. THE INVERSE PROBLEM OF VARIATIONAL CALCULUS

First, we shall formulate a theorem on the solvability of nonlinear equations in scales of Banach spaces which will be used essentially in this section.

Suppose $\mathbf{C}(0,T;X_s)$ is the space of functions analytic in the disk $\mathbf{C}_T = \{z \in \mathbb{C} \mid |z| < T\}$, bounded for $|z| \leq T$ and taking values in the Banach space X_s. Defining the norm

$$\|u\|_{\mathbf{C}(0,T;X_s)} = \sup_{t \in [0,T]} \|u\|_{s,t},$$

$$\|u\|_{s,t} = \sup_{|z|=t} \|u(z)\|_s, \qquad \|\cdot\| = \|\cdot\|_{X_s},$$

we make $\mathbf{C}(0,T;X_s)$ a Banach space.

Here X_s, $s \in [0,1]$ is a one-parameter set (a scale) of Banach spaces such that $X_s \subseteq X_{s'}$ for $s' < s$, where the norm of the embedding operator is less or equal to 1, i.e., for all $u \in X_s$ we have

$$\|u\|_{s'} \leq \|u\|_s, \qquad s' < s.$$

Suppose that for any pair of numbers $s, s' \in [0,1]$, $s' < s$, a mapping V is defined on the ball

$$\mathbf{C}^{r,u_0}(0,T;X_s) = \{u \in \mathbf{C}(0,T;X_s) \mid \|u - u_0\|_{\mathbf{C}(0,T;X_{s'})} < r\}$$

with center $u_0 \in \mathbf{C}(0,T,X_1)$ and maps this ball to $\mathbf{C}(0,T;X_{s'})$. We call V a Volterra operator of class $J(\alpha, \beta, C)$, $\alpha > 0$, $\beta \geq 0$ if there exists a number $c > 0$ such that for any $u,v \in \mathbf{C}^{r,u_0}(0,T;X_s)$, $s' < s, t \in [0,T]$ the following estimate holds:

$$\|Vu - Vv\|_{s',t} \leq c(s-s')^{-\beta}(J^\alpha \|u-v\|_{s,\tau})(t),$$

Here J^α is the operator of integration of order $\alpha > 0$:

$$(J^\alpha \varphi)(t) = \frac{1}{\Gamma(\alpha)} \int_0^t (t-\tau)^{\alpha-1} \varphi(\tau)\, d\tau,$$

and $\Gamma(\alpha)$ is the gamma function. In particular, for $\alpha = 1$ we have

$$(J\varphi)(t) = \int_0^t \varphi(\tau)\, d\tau, \qquad J \equiv J^1.$$

Theorem 4.1. *Suppose $V \in J(\alpha, \alpha, \mathbf{C})$. Then 1) the solution of the equation $u = Vu$ is unique in the ball $\mathbf{C}^{r,u_0}(0,T;X_s)$ for $s > 0$; 2) if $Vu_0 \in \mathbf{C}^{r,u_0}(0,T;X_s)$ for some $s \in (0,1]$, then there exists a number $a > 0$ such that for any $s' < s$ the equation $u = Vu$ has a solution*

$$u \in \mathbf{C}^{r,u_0}(0,T';X_{s'}), \qquad T' < a(s-s'), \qquad T' < T.$$

The proof of this theorem will be given at the end of the section.

Now, we shall state the inverse problem.

The classical problem of variational calculus is as follows. Given a function $L(x,y,t)$, $x,y,t \in \mathbb{R}$, find a curve $y = y(x)$ passing through the points $(x_0, 0)$ and $(x_1, 0)$ such that the value of the integral

$$\tau(x_0, x_1) = \int_{x_0}^{x_1} L(x, y, y')\, dx \tag{4.1}$$

is minimal. In this section, we consider the inverse problem which consists in finding the function L of the form

$$L(x,y,t) = \frac{\sqrt{1+t^2}}{V(x,y)}, \qquad v = ? \tag{4.2}$$

by the function τ. Such problem arises in geophysics (the inverse kinematic seismic problem), in medicine (ultrasonic tomography), and in plasma physics (diagnosis of plasma). The curves for which the functional (4.1) for the Lagrangian L of the form (4.2) achieves its minimum are geodetic lines of the metric

$$ds^2 = v^{-2}(dx^2 + dy^2).$$

The function v is the velocity and τ is the time of wave propagation in the studied medium. The inverse problem is reduced to solution of the nonlinear integral equation

$$\Phi v \equiv \int_{x_0}^{x_1} \frac{\sqrt{1+y'^2(x)}}{v(x,y)}\, dx = \tau(x_0, x_1), \qquad x_0, x_1 \in (-\varepsilon, \varepsilon), \tag{4.3}$$

Chapter 4. Volterra operator equations and their applications   125

where $y = y(x)$ is a solution of the Euler equation

$$y'' = F(v, v_x, v_y, y') \equiv \frac{(1 + y'^2)(y'v_x - v_y)}{v}, \quad (4.4)$$

$$y(x_0) = 0, \qquad y(x_1) > 0. \quad (4.5)$$

We seek a solution v of the equation $\Phi v = \tau$ in the space of smooth functions v such that

$$v(x, y) > 0, \qquad v_y(x, y) > 0. \quad (4.6)$$

This condition provides (locally) the return of the geodetic lines $y = y(x)$ going from the point $(x_0, 0)$ to the point $(x_1, 0)$ and their upward convexity, since for small y', due to (4.6), $y'' = F < 0$. We show that for the solvability of equation (4.3) the following conditions are necessary:

$$\tau(x_0, x_1) + \tau(x_1, x_0) = 0, \quad (4.7)$$

$$\tau_{x_0}(x_0, x_0) < 0, \qquad \tau_{x_0 x_1 x_1}(x_0, x_0) > 0. \quad (4.8)$$

Condition (4.7) follows from equation (4.3) and the fact that

$$y = y(x, x_0, x_1) = y(x, x_1, x_0)$$

(see (4.4) and (4.5)). To prove inequality (4.8), we recall the formula of differentiation of the function τ from variational calculus:

$$\tau_{x_0}(x_0, x_1) = L(x_0, 0, t_0) + t_0 L'_{t_0}(x_0, 0, t_0), \quad (4.9)$$

$$t_0 = y'(x_0). \quad (4.10)$$

By the Taylor formula, we have

$$y(x) = y(x_0) + y'(x_0)(x - x_0) + y''(x_0)\frac{(x - x_0)^2}{2} + O((x - x_0)^3). \quad (4.11)$$

Setting $x = x_1$ in this formula and taking into account (4.5) and (4.10), we obtain

$$t_0 = O(x_1 - x_0). \quad (4.12)$$

Since $t_0 \to 0$ as $x_1 \to x_0$, from (4.2), (4.6), and (4.9) it follows that

$$\tau_{x_0}(x_0, x_0) = -v^{-1}(x_0, 0) < 0.$$

Formulas (4.4) and (4.12) yield

$$y''(x_0) = -v_y(x_0,0)/v(x_0,0) + O(x_1 - x_0). \tag{4.13}$$

Thus, in view of (4.13), formula (4.11) with $x = x_1$ yields

$$t_0 = \frac{v_y(x_0,0)}{v(x_0,0)} \frac{(x_1 - x_0)}{2} + O((x_1 - x_0)^2). \tag{4.14}$$

Since

$$(1 + t_0^2)^{\pm 1/2} = 1 \pm t_0^2/2 + O(t_0^4),$$

from (4.2), (4.9), and (4.14) we obtain

$$\tau_{x_0}(x_0, x_1) = v^{-1}(x_0, 0) \left\{ -1 + [v^{-1}(x_0, 0)v_y(x_0, 0)]^2 \frac{(x_1 - x_0)^2}{8} \right\}$$
$$+ O((x_1 - x_0)^3).$$

Hence, $\tau_{x_0 x_1 x_1}(x_0, x_0) > 0$ and estimates (4.8) are established. We now prove that in the class C^ω of real analytic functions these conditions are sufficient for local solvability.

Theorem 4.2. *Suppose $\mathcal{Y} = (-\varepsilon, \varepsilon) \times (-\varepsilon, \varepsilon)$, $\tau \in C^\omega(\mathcal{Y})$ and τ satisfies conditions (4.7), (4.8) in $\mathcal{Y}$. Then there exists a unique function v defined in some neighbourhood of zero, such that $v \in C^\omega$, $v > 0$, $v_y > 0$, and $\Phi v = \tau$ in this neighbourhood.*

Proof. We construct the function

$$T(x_0, x_1) = \tan \arcsin \left\{ (x_1 - x_0) \left[\frac{\tau_{x_0}^2(x_0, x_0) - \tau_{x_0}^2(x_0, x_1)}{(x_1 - x_0)^2} \right]^{1/2} \right\}. \tag{4.15}$$

Applying the L'Hospital rule twice and using the equality

$$\tau_{x_0 x_1}(x_0, x_0) = D_{x_0 x_1}(\tau(x_0, x_0) + \tau(x_1, x_0))/2|_{x_1 = x_0} = 0,$$

we see that the limit of the function in square brackets as $x_1 \to x_0$ is equal to

$$-\tau_{x_0}(x_0, x_0) \tau_{x_0 x_1 x_1}(x_0, x_0) > 0.$$

From (4.15) it follows that

$$-T'_{x_0}(x_0, x_0) = T'_{x_1}(x_0, x_0) = \sqrt{-\tau_{x_0}(x_0, x_0) \tau_{x_0 x_1 x_1}(x_0, x_0)} > 0. \tag{4.16}$$

Chapter 4. Volterra operator equations and their applications

Thus, for sufficiently small $\varepsilon > 0$ we have $T \in C^\omega(\mathcal{Y})$, and by the implicit function theorem there exist neighbourhoods $U_1 \subset \mathbb{R}^2$ and $U_2 \subset \mathbb{R}^1$ of zero, such that for all $(a_0, t_0) \in U_1$ the equation $T(x_0, x_1) = t_0$ has a unique solution $x_1 = \varphi_1(x_0, t_0) \in U_2$, where $\varphi_1 \in C^\omega(U_1)$. We set

$$\varphi_2(x_0, t_0) = T(\varphi_1(x_0, t_0), x_0), \qquad \varphi = (\varphi_1, \varphi_2) \qquad (4.17)$$

$$U = U_n \cup \varphi(U_n), \qquad U_n = (1/n)U_1,$$

where the number n is chosen so that $\varphi(U) \subseteq U_1$. We show that U is an open set, $\varphi(U) = U$, and

$$\varphi^2 = \varphi \circ \varphi = E, \qquad (x_0, t_0) \in U, \qquad (4.18)$$

where E is the identity mapping. Indeed, by the definition of φ_1 we have

$$T(x_0, \varphi_1(x_0, t_0)) = t_0. \qquad (4.19)$$

Substituting the function $\varphi_1(x_0, t_0)$ for x_0 and the function $\varphi_2(x_0, t_0)$ for t_0 in this identity, from (4.17) and (4.19) we obtain

$$T(\varphi_1(x_0, t_0), \varphi_1(\varphi_1(x_0, t_0), \varphi_2(x_0, t_0))) = T(\varphi_1(x_0, t_0), x_0).$$

Hence, by the local solvability of the equation $T(x_0, x_1) = t_0$, we have

$$\varphi_1(\varphi_1(x_0, t_0), \varphi_2(x_0, t_0)) = x_0. \qquad (4.20)$$

Analogously, from the definition of φ_2 and from (4.19) and (4.20) we find

$$\varphi_2(\varphi_1(x_0, t_0), \varphi_2(x_0, t_0)) = T(\varphi_1(\varphi_1(x_0, t_0), \varphi_2(x_0, t_0))),$$

$$\varphi_1(x_0, t_0) = T(x_0, \varphi_1(x_0, t_0)) = t_0.$$

Thus, identity (4.18) is proved. From this identity it follows that

$$\varphi(U) = \varphi(U_n) \cup \varphi^2(U_n) = \varphi(U_n) \cup U_n = U.$$

Since $\varphi(U_n) = \varphi^{-1}(U_n)$, the set U is open because U_n is open and φ is continuous. Now we set

$$f(x_0, t_0) = \tau(x_0, \varphi_1(x_0, t_0)), \quad g(x_0) = -\tau_{x_0}^{-1}(x_0, x_0), \qquad (4.21)$$

$$f_\pm = (f + f \circ \varphi)/2.$$

Formulas (4.7), (4.8), (4.16), and (4.19) yield

$$f'_{x_0}(x_0,0) = 0, \qquad f'_{t_0}(x_0,0) > 0, \qquad g(x_0) > 0. \qquad (4.22)$$

Moreover, from (4.7) and (4.20) we have $f_\pm(x_0,t_0) \equiv 0$, i.e., $f = f_-$ is an odd function with respect to the mapping φ.

Consider the problem of determination of the functions $u(x,y,t)$ and $v(x,y)$ from the conditions

$$tu_y + u_x + F(x, v_x, v_y, t)u_t + 2L(v,t) = 0, \qquad (4.23)$$

$$u(x,0,t) = f(x,t), \qquad v(x,0) = g(x), \qquad (x,t) \in U. \qquad (4.24)$$

Here the functions L, F, t, and g are given by formulas (4.2), (4.4), and (4.12). This is an unusual problem because two functions are to be determined from one equation and the Cauchy data for u are given in the plane, which is characteristic for $t = 0$.

Therefore, we need no additional condition for v, as it is obtained from (4.23) for $t = 0$:

$$v_y = \frac{u_x(x,y,0)}{u_t(x,y,0)} v(x,y) + \frac{2}{u_t(x,y,0)}. \qquad (4.25)$$

Applying the difference operator

$$\Delta u = (U(x,y,t) - u(x,y,0))/t$$

to equation (4.23), we obtain

$$u_y = -\Delta\{u_x + F(v, v_x, v_y, t)u_t + 2L(v,t)\}.$$

Using equation (4.25) to eliminate the function v_y in this equation and integrating it together with equation (4.25) with respect to y, in view of (4.24), (4.2), and (4.4) we obtain the system of equations

$$u(x,y,t)$$
$$= -\int_0^y \Delta\left\{u_x(x,\eta,t)\frac{1+t^2}{v(x,\eta)}\left[tv_x - \frac{u_x(x,\eta,0)}{u_t(x,\eta,0)}v(x,\eta) - \frac{2}{u_t(x,\eta,0)}\right]\right.$$
$$\left. + \frac{2(1+t)^{1/2}}{v(x,\eta)}\right\} d\eta + f(x,t), \qquad (4.26)$$

$$v(x,y) = \int_0^y \left[\frac{u_x(x,\eta,0)}{u_t(x,\eta,0)}v(x,\eta) + \frac{2}{u_t(x,\eta,0)}\right] d\eta + g(x), \qquad (4.27)$$

Chapter 4. Volterra operator equations and their applications

which is equivalent to the system (4.23), (2.24).

Introducing the vector function $w = (u, v)$, we rewrite this system in the form $w = Vw$, where the operator V is defined by the right-hand sides of equalities (4.26) and (4.27). Suppose the Banach space X_s, $s \in [0, 1]$ consists of the vector functions $w : \Omega_s \to \mathbb{C}^2$ analytic in the domain

$$\Omega_s = \{x, t \in \mathbb{C} \mid |x| < \delta(1+s), \quad |t| < \delta(1+s)\}$$

such that

$$\|w\|_s = |w|_s + |w_x|_s + |w_t|_s < \infty. \tag{4.28}$$

Here $|w|_s = \sup |w(x,t)|$, $(x,t) \in \Omega_s$. We set $w_0 = (f(x,t), g(s))$ and show that the operator V satisfies the assumptions of Theorem 4.1, where $|y|$ stands for t. For sufficiently small δ, the vector w belongs to $C(0, T; X_1)$ and there exists a number $r > 0$ such that

$$|f_t(x, t)| \geq 2r, \quad |g(x)| \geq 2r$$

for all $(x, t) \in \Omega_1$ (see (4.22)). For $w = (u, v) \in \mathbf{C}^{r, w_0}(0, T; X_s)$ we have

$$|u_t|_s \geq |f_t|_s - |u_t - f_t|_s \geq 2r - r = r,$$
$$|v|_s \geq |g|_s - |v - g|_s \geq r.$$

Therefore, division by the functions u_t and v in (4.26) and (4.27) is correct and

$$V : \mathbf{C}^{r, w_0}(0, T; X_s) \to C(0, T; X_{s'}), \qquad s' < s.$$

From the definition of the mapping V it follows that for $w_1, w_2 \in \mathbf{C}^{r, w_0}(0, T; X_s)$ the difference $V w_1 - V w_2$ is represented in the form of the linear combination of the vectors $w_1 - w_2$, $D(w_1 - w_2)$, $D(w_1 - w_2)|_{t=0}$ ($D = D_x$ or $D = D_t$) with operator coefficients of the form $J\Delta^k a$, $k = 0, 1$. Here the function a is expressed in terms of w_1, w_2, and their first derivatives; J is the operator of integration from 0 to y. Note that

$$|Ju|_s \leq J|u|_s, \qquad |au|_s \leq |a|_s |u|_s$$

$$|Du|_{s'} \leq \delta^{-1}(s - s')^{-1}|u|_{s'}, \qquad s' < s,$$

$$|\Delta u|_s \leq 2\delta^{-1}|u|_s. \tag{4.29}$$

The first two estimates are trivial, and the third one follows from the Cauchy formula for analytic functions. The estimate (4.29) follows from the fact that an analytic function achieves its maximum on the boundary, i.e., in our case, for $|t| = \delta(1+s) \geq \delta$. Thus, taking into account (4.28), we obtain

$$|D^\alpha(Vw_1 - Vw_2)|_{s'} \leq c(s-s')^{-|\alpha|} J\|w_1 - w_2\|_s$$

$$\|Vw_1 - Vw_2\|_{s'} \leq 3c(s-s')^{-1} J\|w_1 - w_2\|_s.$$

These estimates are obtained for real y. Evidently, they remain valid for complex y if we take the operator of integration from O to $|y|$ for J in the right-hand sides of the estimates. The obtained estimates show that $V \in J(1, 1, \mathbf{C})$. If $|y| < T$, then from the definition of V we have

$$\|Vw_0 - w_0\|_s = O(T) < r$$

for sufficiently small T. Thus, Theorem 4.1 implies the existence of an analytic solution $w = (u(x, y, t), v(x, y))$ of the system (4.26), (4.27) and, consequently, of the problem (4.23), (4.24) in a certain neighbourhood of zero. From conditions (4.22), (4.24), and equation (4.25) for $y = 0$ it follows that $v > 0$ and $v_y > 0$ in a real neighbourhood of zero. Suppose $y = y(x)$ is a solution of the boundary value problem (4.4), (4.5), where $x_1 = \varphi_1(x_0, t_0)$. The existence of $y(x)$ for small t_0 follows from the fact that $\varphi(x_0, t_0) \to x_0$ for $t_0 \to 0$ and $F < 0$ for small y'. From equation (4.23) and the definition of functions F and L (see (4.2), (4.4)) it follows that

$$\frac{du(x, y, y')}{dx} = -2\frac{(1+y'^2)^{1/2}}{v(x, y)}$$

along the curve $y(x)$. Integrating this identity with respect to x from x_0 to $\varphi_1(x_0, t_0)$, we have

$$\int_{x_0}^{\varphi_1(x_0, t_0)} (1+y'^2)^{1/2} \frac{dx}{v(x, y)} = f_-(x_0, t_0) = f = \tau(x_0, \varphi_1(x_0, t_0))$$

because $u(x, 0, t) = f(x, t)$ and $f_- = f$. Setting $x_1 = \varphi_1(x_0, t_0)$, we obtain the equality $\Phi v = \tau$, which is the required result. $\square$

Remark 4.1. Suppose $y(x)$ is a solution of equation (4.4) with the data $y(x_0) = 0$, $y'(x_0) = t_0$. Then from formulas (4.2) and (4.9) it follows that $\varphi_1(x_0, t_0)$ and $\varphi_2(x_0, t_0)$ are the second end $x_1 = \varphi_1$ and the tangent of the slope of the curve $y(x)$ at this point, respectively, i.e., $y(\varphi_1) = 0$, $y'(\varphi_1) = \varphi_2$. This is the geometrical sense of the mapping φ. It is also clear that the function $T(x_0, x_1)$ defined by (4.15) represents the tangent

Chapter 4. Volterra operator equations and their applications

of the slope of the curve $y(x)$, which is a solution, in terms of the function $T(x_0, x_1)$, of the boundary value problem (4.4), (4.5). This clarifies a somewhat strange appearance of the function $T(x_0, x_1)$ in the beginning of the proof of Theorem 4.2.

Exercise 4.1. For positive numbers a and α we denote by $B^\alpha(a)$ the Banach space of analytic functions $u(z)$ with values in X_s for $|z| = t < a(1-s)$. The norm $N[u]$ in $B^\alpha(a)$ is defined as follows:

$$N[u] = \sup(\rho^\alpha(t,s)\|u\|_{s,t}).$$

Here the supremum is taken over the set

$$\Delta_a = \{t, s \mid s \in (0,1),\quad 0 < t < a(1-s)\},$$

$a > 0$ is a number, $\rho(t,s) = a(1-s)t^{-1} - 1$ is a weight, and the norm $\|u\|_{s,t}$ is defined in the beginning of the section. Suppose that for a function $w(z)$ with values in the space X_s for $|z| = t < a(1-\delta)$ and for a function $v \in B^\alpha(a)$ the inequality

$$\|w\|_{s',t} \le c(s-s')^{-\alpha} \int_0^t (t-\tau)^{\alpha-1} \|v\|_{s,\tau}\,d\tau$$

holds for all $(t,s) \in \Delta_a$ and $s' \in [0,s)$. Prove the estimate $N[w] \le O(a^\alpha) N[v]$. Show that if

$$\|w\|_{s',t} \le c(1-s)^{-\alpha} \int_0^t (t-\tau)^{\alpha-1} \|v\|_{1,\tau}\,d\tau,$$

for all $(t,s) \in \Delta_a$, where the function v belongs to a ball of fixed radius in the space $\mathbf{C}(0, x; X_1)$, then $N[w] = O(a^\alpha)$ uniformly with respect to V.

Exercise 4.2. Suppose $a_0 > 0$, $a_k(1+(k+1)^{-2})^{-1}$, $k = 0, 1, 2, \ldots$. Denote the norm in the space $B^\alpha(a_k)$ by $N_k[u]$. Prove the following statements:
 (i) $a = \lim a_k > 0$, $N_{k+1}[u] \le N_k[u]$;
 (ii) if $t \le a_{k+1}(1-s)$ then $\|u\|_{\mathbf{C}(0,t;X_s)} \le (k+1)^{2\alpha} N_k[u]$.

Exercise 4.3. Suppose that $V \in J(\alpha, \alpha, \mathbf{C})$, $V(u_0) \in \mathbf{C}^{r,u_0}(0, T; X_1)$. Set

$$v_0 = 0,\quad v_{k+1} = V(v_k + V(u_0)) - V(u_0),\quad w_k = v_{k+1} - v_k,\quad \lambda_k = N_k[w_k].$$

Here $N_k[u]$ is the norm in the space $B^\alpha(a_k)$ from Exercise 4.2. Prove the following statements:

(i) $\lambda_k \leq O(a_0^\alpha)\lambda_{k-1}$;

(ii) for sufficiently small a_0, there exists a number $r' < r - \|V(u_0) - u_0\|_{C(0,T;X_1)}$ such that the sequence v_k converges in the space $B^\alpha(a)$, where $a = \lim a_k$, to a certain function v and $\|v\|_{C(0,t;X_s)} < r'$ for $t < a(1-s)$;

(iii) the function $u = v + V(u_0) \in C^{r,u_0}(0,t;X_s)$ and $u = V(u)$ for $|z| \leq t$.

The last statement proves the assertion of Theorem 4.1 concerning the existence of a solution for $s = 1$.

The general case is established in a similar way. The proof of the uniqueness of the solution is similar to that in Theorem 1.1.

Exercise 4.4. Prove formula (4.9).

4.5. REMARKS AND REFERENCES

Basic results of Section 1 are contained in Bukhgeim (1983) in more general form. Theorem 1.3 was proved by Freeman (1965). Proving this theorem, we followed Dunford and Schwartz (1971). The theory of Volterra operator equations is an essential generalization of the theory of abstract Cauchy problem in scales of Banach spaces, which is developed by Ovsyannikov (1965, 1971), Nirenberg (1977), Treves (1967), and other authors (see the bibliography therein).

The results of Sections 2 and 3 are published for the first time in detail.

The results of Section 4 are contained in the brief form in Bukhgeim (1983).

Chapter 5.

Foundations of the theory of conditionally well-posed problems

5.1. CONDITIONAL WELL-POSEDNESS

Suppose A is a linear operator acting from a normed space X to a normed space Y, with domain of definition $D(A) \subseteq X$ and range of values $R(A) \subseteq Y$,

$$A : D(A) \to Y. \tag{1.1}$$

We consider the problem of solving the equation

$$Au = f, \tag{1.2}$$

where f is a given element of the space Y.

Definition 1.1. Problem (1.2) is said to be *well-posed* if

$$\|u\| \leq c\|Au\| \tag{1.3}$$

for all $u \in D(A)$, where the number c is independent of u and $\|\cdot\|$ is a norm in the space X or Y. Problem (1.2) is said to be ill-posed if estimate (1.3) does not hold for it.

Well-posedness of the problem (1.2) guarantees the uniqueness of the solution u, i.e.,

$$\operatorname{Ker} A = \{v \in D(A) \mid Av = 0\} = \{0\},$$

and stability:
$$\|f\| \leq \varepsilon \Rightarrow \|u\| \leq c\varepsilon.$$

If we assume in addition that the spaces X and Y are Banach spaces, the range of values $R(A) = AD(A)$ is dense in the space Y, and A is a closed operator, then we may state that for any right-hand side $f \in Y$ there exists a solution u of the equation $Au = f$ (see Exercise 1.1). The presence of these three properties (uniqueness, stability, and existence of a solution) is often taken as the definition of a well-posed problem. It is clear that well-posedness of the problem depends essentially on the spaces X and Y, in which it is considered, and also on the domain of definition $D(A)$. In applications, the choice of functional spaces X and Y is usually dictated by the specific characteristics of the problem. However, the definition of the norm in X or Y involves no more than a finite number of derivatives. This fact is concerned with real characteristics of measuring devices. The following examples of functional spaces are most typical for applications.

Suppose Ω is a domain in $\mathbb{R}^n$. We denote by $C^m(\overline{\Omega})$ the space of functions $u(x)$ defined in Ω, m times continuously differentiable, whose derivatives $\partial^\alpha u(x)$, $|\alpha| \leq m$ can be extended to $\overline{\Omega}$ continuously. The norm in $C^m(\overline{\Omega})$ is defined as follows:

$$\|u\|_{C^m(\overline{\Omega})} = \sup_{\substack{|\alpha| \leq m \\ x \in \Omega}} |\partial^\alpha u(x)|. \tag{1.4}$$

Here, as usual, $\alpha = (\alpha_1, \alpha_2, \ldots, \alpha_n)$ is a multi-index,

$$\partial^\alpha = \partial_1^{\alpha_1} \ldots \partial_n^{\alpha_n}, \qquad \partial_j = \partial x_j = \partial/\partial x_j, \qquad |\alpha| = \alpha_1 + \alpha_2 + \ldots + \alpha_n.$$

The space $L_p(\Omega)$ is defined as the set of all functions $u(x)$ for which $|u(x)|^p$ is integrable over Ω and

$$\|u\|_p = \left\{ \int_\Omega |u(x)|^p \, dx \right\}^{1/p}, \qquad 1 \leq p \leq \infty.$$

For $p = \infty$, we define
$$\|u\|_\infty = \operatorname*{vrai\,sup}_{x \in \Omega} |u(x)|.$$

The norm in the Sobolev space $W_p^m(\Omega)$ is defined by the formula

$$\|u\|_{m,p} = \left\{ \sum_{\alpha \leq m} \|D^\alpha u\|_p^p \right\}^{1/p}, \qquad 1 \leq p \leq \infty.$$

Now, we consider elementary examples of the problem (1.1)–(1.2).

Chapter 5. Theory of conditionally well-posed problems

Example 1.1. Let $X = Y = C[0,T]$,
$$Au = du/ddt, \qquad D(A) = C^1[0,T].$$

Evidently, the problem of solution of the equation $Au = f$ is ill-posed since $Au = 0$ for $u(t) \equiv 1$. The problem becomes well-posed if we narrow down the domain of definition of the operator A to the set $D_0(A) = \{u \in C^1[0,T] \mid u(0) = 0\}$. In this case,

$$u(t) = \int_0^t f(\tau)\, d\tau = \int_0^t (Au)(\tau)\, d\tau, \qquad t \in [0,T],$$

and, consequently,

$$\|u\| \le T\|Au\|, \qquad \forall u \in D_0(A).$$

Example 1.2. Suppose $X = Y = C[0,T]$,

$$(Au)(t) = \int_0^t u(\tau)\, d\tau, \qquad t \in [0,T], \quad D(A) = C[0,T].$$

We show that this problem is ill-posed. Set $u_n(t) = n \cos nt$. Then

$$Au_n = \int_0^t n \cos n\tau\, d\tau = \sin nt$$

and
$$\|Au_n\| = 1, \qquad \|u_n\| = n. \tag{1.5}$$

Substituting formulas (1.5) into estimate (1.3) and passing to the limit as $n \to \infty$, we arrive at the contradiction with the assumption of well-posedness of the problem $Au = f$. Now, in this example we take $C^1[0,T]$ for the space Y. Then from the definition of the norm (1.4) it follows that $\|Au\| \ge \|u\|$ and the problem becomes well-posed.

Example 1.3. Suppose that the function $v(x,t)$ belongs to the space $C^2(\overline{\Omega})$, where $\overline{\Omega} = [0,\pi] \times [0,T]$, and is a solution of the boundary value problem

$$v_t = v_{xx}, \qquad (x,t) \in \Omega, \tag{1.6}$$
$$v(x,0) = f(x), \qquad x \in [0,\pi], \tag{1.7}$$
$$v(0,t) = v(\pi,t) = 0, \qquad t \in [0,T], \tag{1.8}$$

where f is a given function from the space $C^2[0,T]$. Set $u(x) = v(x,T)$. We define the operator A by the rule $A : u(x) \to f(x)$, i.e.,

$$(Au)(x) = f(x), \quad x \in [0, \pi].$$

We prove that

$$\int_0^\pi |u(x)|^2 \, dx \leq \int_0^\pi |f(x)|^2 \, dx, \tag{1.9}$$

i.e., in the space $L_2(0,\pi)$

$$\|u\| \leq \|Au\|, \quad \forall u \in D(A).$$

Note that the domain of definition of the operator A in this example is given implicitly as a set of functions $u(x)$ which admit the representation of the form $u(x) = v(x,T)$. Here v is a solution of the boundary value problem (1.6)–(1.8). For both X and Y we take the functional space $L_2(0,\pi)$.

To establish estimate (1.9), we set

$$E(t) = \int_0^\pi v^2(x,t) \, dx$$

and show that

$$E(t) \leq E(0), \quad \forall t \in [0,T]. \tag{1.10}$$

Differentiating the function $E(t)$ and taking conditions (1.6)–(1.8) into account, we have

$$E'(t) = 2 \int_0^\pi v(x,t) v_t(x,t) \, dx = 2 \int_0^\pi v(x,t) v_{xx}(x,t) \, dx$$

$$= -2 \int_0^\pi v_x^2(x,t) \, dx \leq 0,$$

which implies estimate (1.10). Setting $t = T$ in this estimate, we obtain estimate (1.9). Thus, the problem $Au = f$ is well-posed.

Example 1.4. Now, we suppose that A is defined as in Example 1.3, but the function $v(x,t)$ satisfies the equation

$$v_t = -v_{xx}$$

instead of (1.6).

We show that the problem $Au = f$ is ill-posed. To this end, we set $f_n(x) = \sin nx$. It is easy to see that

$$v(x,t) = \exp(n^2 t) \sin nx = \exp(n^2 t) f_n(x)$$

and therefore

$$u_n(x) = v(x,T) = \exp(n^2 T) f_n(x).$$

Substituting u_n, $Au_n = f_n$ into estimate (1.3) and dividing by $\|f_n\| \neq 0$, we obtain $\exp(n^2 T) \leq C$. As $n \to \infty$, we arrive at a contradiction.

For many linear ill-posed problems of the form

$$Au = f, \quad u \in D(A) \subset X, \quad A: D(A) \to Y$$

some additional information about the solution u is known, namely that $u \in D(l)$, where l is a functional defined in some linear space $D(l)$ with the properties of seminorm. This means that

$$l(u) \geq 0, \quad l(\lambda u) = |\lambda| l(u), \quad l(u+v) \quad l(u) + l(v).$$

Here $u, v \in D(l)$, $\lambda \in \mathbb{R}$ if the space is real, and $\lambda \in \mathbb{C}$ if $D(l)$ is complex. Moreover, $D(l) \cap X \neq \emptyset$. It is natural to give the following generalization of Definition 1.1.

Definition 1.2. The problem $Au = f$ is called *l-well-posed* if for any $\varepsilon > 0$ there exists a positive constant $c(\varepsilon)$ such that for all $u \in D(A) \cap D(l)$ the following estimate holds:

$$\|u\| \leq \varepsilon l(u) + c(\varepsilon) \|Au\|. \tag{1.11}$$

The functional l of the l-well-posed problem $Au = f$ is called *the stabilizing functional*.

Following this definition, well-posed problems become 0-well posed, where $c(\varepsilon)$ does not depend on ε. If an l-well-posed problem is not well-posed in $D(A) \cap D(l)$, then

$$\lim_{\varepsilon \to 0} c(\varepsilon) = \infty.$$

Indeed, otherwise, if we pass to the limit as $\varepsilon \to 0$ in (1.11), we obtain estimate (1.3). We also note that the stabilizing functional may be considered strictly positive without loss of generality. More precisely,

$$\forall u \in D(l) \cap X, \quad l(u) \geq c_0 \|u\|_X, \quad c_0 > 0, \tag{1.12}$$

otherwise the functional can be replaced by
$$l_1(u) = l(u) + c_0\|u\|_X.$$

We show that the function $c(\varepsilon)$ from estimate (1.11) may be always considered as a nonincreasing function of the parameter ε. Indeed, for any $0 < \varepsilon' \leq \varepsilon$ from (1.11) we have
$$\|u\| \leq \varepsilon' l(u) + c(\varepsilon')\|Au\| \leq \varepsilon l(u) + c(\varepsilon')\|Au\|.$$

Therefore, setting
$$c_1(\varepsilon) = \inf_{0<\varepsilon'\leq \varepsilon} c(\varepsilon'), \tag{1.13}$$

we have $\|u\| \leq \varepsilon l(u) + c_1(\varepsilon)\|Au\|$. On the other hand, from definition (1.13) it follows that $c_1(\varepsilon_1) \geq c_1(\varepsilon_2)$ for $\varepsilon_1 < \varepsilon_2$, i.e., $c_1(\varepsilon)$ is a nonincreasing function for $\varepsilon > 0$.

The l-well-posedness of problem (1.2) guarantees the uniqueness of its solution in $D(A) \cap D(l)$ and stability in each set M of the form
$$M = \{u \in D(A) \cap D(l) \mid l(u) \leq m\}.$$

To show the last property, it suffices to take infimum with respect to ε of the right-hand side of inequality (1.11) and to take into account the condition $l(u) \leq m$. Then we obtain
$$\|u\| \leq \omega_m(\|Au\|), \qquad \omega_m(\delta) = \inf_{\varepsilon>0}(\varepsilon m + c(\varepsilon)\delta).$$

Evidently, $\omega_m(\delta)$ is continuous at the zero point and $\omega_m(0) = 0$.

The following examples are elementary examples of l-well-posed problems.

Example 1.5. Suppose that $A \in \mathcal{L}(X, Y)$, i.e., A is a continuous operator from X to Y. Suppose $D(A) = X$ is a Banach space and l is an a seminorm defined in $D(l)$ such that the set $\{u \in D(l) \cap X \mid l(u) \leq 1\}$ is a compact set in X and estimate (1.12) holds. We show that the problem $Au = f$ is l-well-posed if $\text{Ker } A = \{0\}$. Indeed, suppose that estimate (1.11) is not valid. Then there exists $\varepsilon > 0$ such that for any natural n there exists an element $u_n \in D(l) \cap X$ such that
$$\|u_n\| > \varepsilon l(u_n) + n\|Au_n\|. \tag{1.14}$$

Chapter 5. Theory of conditionally well-posed problems

Estimates (1.14) and (1.12) yield $u_n \neq 0$, $l(u_n) \neq 0$. Set $v_n = u_n/l(u_n)$. Then

$$1 = l(v_n) \geq c_0\|v_n\|, \tag{1.15}$$

$$\|v_n\| > \varepsilon + n\|Av_n\|. \tag{1.16}$$

These estimates yield

$$\|Av_n\| < n^{-1}\|v_n\| - n^{-1}\varepsilon \leq n^{-1}/c_0 - n^{-1}\varepsilon \to 0. \tag{1.17}$$

Since the sequence v_n is compact in X, without loss of generality we may assume that v_n converges in X to a certain element $v \in X$. On the other hand, $A \in \mathcal{L}(X, Y)$ and therefore

$$\|Av_n - Av\| \leq c\|v_n - v\| \to 0. \tag{1.18}$$

Comparing (1.16), (1.17), and (1.18), we have

$$Av = 0, \qquad \|v\| \geq \varepsilon,$$

which contradicts the equality $\operatorname{Ker} A = \{0\}$.

Example 1.6. Suppose $X = Y$ is a Hilbert space, $A \in \mathcal{L}(X, X) = \mathcal{L}(X)$, $\operatorname{Ker} A = \{0\}$, and the range of values $R(A) = AX$ of the operator A is dense everywhere in X. Hence, the operator $A^{-1} : R(A) \to X$ is defined on a manifold which is dense in X and therefore the conjugate operator $(A^{-1})^*$ is defined correctly. Let

$$D(l) = D((A^{-1})^*), \qquad l(u) = \|(A^{-1})^* u\|.$$

We show that the problem $Au = f$ is l-well-posed and

$$\|u\| \leq \varepsilon l(u) + (2\varepsilon)^{-1}\|Au\|, \tag{1.19}$$

i.e., $c(\varepsilon) = 1/2\varepsilon$. Indeed, if $\langle u, v\rangle$ is a scalar product in X, then

$$\|u\|^2 = \langle u, u\rangle = \langle u, A^{-1}Au\rangle = \langle (A^{-1})^* u, Au\rangle$$
$$\leq \|(A^{-1})^*\|\|Au\| \leq \varepsilon^2 \|(A^{-1})^* u\|^2 + (2\varepsilon)^{-2}\|Au\|^2.$$

Here we have used the Cauchy–Schwarz–Bunyakovskii inequality and the trivial estimate

$$2ab \leq \alpha a^2 + \alpha^{-1} b^2, \qquad \varepsilon^2 = \alpha/2.$$

Extracting the square root and using the inequality

$$(a^2 + b^2)^{1/2} \leq |a| + |b|, \qquad a, b \geq 0$$

we obtain (1.19).

We use this general construction to obtain the estimate (1.11) of l-stability of the problem of differentiation

$$Au = \int_0^t u(\tau) \, d\tau = f(t), \qquad u \in L_2(0,T) \cap C^1[0,T].$$

We have

$$\|u\|_2^2 = \langle u, \partial_t Au \rangle = \int_0^T u(t) \partial_t (Au)(t) \, dt$$

$$= u(T)(Au)(T) - \int_0^T u'(t)(Au)(t) \, dt$$

$$\leq \varepsilon^2 |u(T)|^2 + (2\varepsilon)^{-2} |(Au)(T)|^2 + \varepsilon^2 \|u'\|_2^2 + (2\varepsilon)^{-2} \|Au\|_2^2.$$

Set $l^2(u) = \|u'\|_2^2 + |u(T)|^2$, $D(l) = C^1[0,T]$, $\|Au\|^2 = \|Au\|_2^2 + |(Au)(T)|^2$.
Then

$$\|u\|_2 \leq \varepsilon l(u) + (2\varepsilon)^{-1} \|Au\|.$$

In particular, if

$$u(T) = 0, \qquad \int_0^T u(t) \, dt = 0, \qquad (1.20)$$

then $\|u\|_2 \leq \varepsilon \|u'\|_2 + (2\varepsilon)^{-1} \|Au\|_2$. Note that after the change of variables

$$v(t) = u(t) - a - bt, \qquad g(t) = f(t) - at - bt^2/2,$$

where the parameters a and b are determined from the conditions

$$f'(T) - a - bT = 0, \qquad f(T) - aT - bT^2/2 = 0,$$

the equation $Au = f$ is transformed into the equation $Av = g$, where the function v already satisfies conditions (1.20).

The disadvantage of estimate (1.19) is that the condition $u \in D(l)$ is difficult to verify in practice because in this case it is necessary to know the structure of the manifold $D((A^{-1})^*)$.

Example 1.7. We consider the problem of Example 1.4:

$$v_t = -v_{xx}, \qquad (x,t) \in \Omega = (0,\pi) \times (0,T), \quad v \in C^2(\overline{\Omega}), \qquad (1.21)$$

$$v(x,0) = f(x), \qquad x \in (0,\pi),$$
$$v(0,t) = v(\pi,t) = 0, \qquad t \in (0,T). \qquad (1.22)$$

Given the function $f(x)$, it is required to find the function $u(x) = v(x,T)$ and obtain the estimate of l-stability for it.

To this end, we consider the integral

$$I = \int_0^T \int_0^\pi \exp(-2st)|v_t + v_{xx}|^2 \, dx \, dt, \qquad s > 0,$$

which is equal to zero, as it follows from equation (1.21). We transform this integral making the change of variables

$$v = \exp(st)w(x,t). \qquad (1.23)$$

We have

$$0 = I = \int_0^T \int_0^\pi |w_t + sw + w_{xx}|^2 \, dx \, dt$$

$$= \int_0^T \int_0^\pi |w_t|^2 \, dx \, dt + \int_0^T \int_0^\pi |sw + w_{xx}|^2 \, dx \, dt$$

$$+ 2 \int_0^T \int_0^\pi (sw_t w + w_t w_{xx}) \, dx \, dt$$

$$\geq \int_0^T \int_0^\pi s\partial_t w^2(x,t) \, dx \, dt - 2 \int_0^T \int_0^\pi w_{tx} w_x \, dx \, dt$$

$$= \int_0^T \int_0^\pi \partial_t(sw^2 - w_x^2) \, dx \, dt = \int_0^\pi (sw^2(x,t) - w_x^2(x,t)) \, dx \Big|_0^T.$$

To obtain this inequality we neglected the first two nonnegative terms in the integral I and integrated by parts with respect to x. By formula (1.23) and the homogeneous boundary conditions (1.22), we have $w_t(0,t) = w_t(\pi,t) = 0$, so, after integrating by parts there appear only integral terms. The last estimate yields

$$s\int_0^\pi w^2(x,T) \, dx - \int_0^\pi w_x^2(x,T) \, dx$$
$$\leq s\int_0^\pi w^2(x,0) \, dx - \int_0^\pi w_x^2(x,0) \, dx \leq s\int_0^\pi w^2(x,0) \, dx.$$

Recalling that

$$w(x,T) = \exp(-sT)v(x,T) = \exp(-sT)u(x), \qquad w(x,0) = v(x,0) = f(x),$$

we obtain

$$s\exp(-2sT)\int_0^\pi |u(x)|^2\,dx \le \exp(-2sT)\int_0^\pi |u'_x(x)|^2\,dx + s\int_0^\pi |f(x)|^2\,dx.$$

Dividing both sides of this inequality by $s\exp(-2sT)$ and setting $\varepsilon^2 = s^{-1}$, after extracting the square root we obtain

$$\|u\|_2 \le \varepsilon\|u_x\|_2 + \exp(T/\varepsilon^2)\|f\|_2.$$

Thus, the problem of solution of the equation $Au = f$ is l-well-posed and $l(u) = \|u_x\|_2$.

Nowadays, the method used in this example is the most powerful method of obtaining uniqueness theorems and the estimates of conditional stability for ill-posed boundary value problems of mathematical physics. The basic idea of this method consists in using integration by parts for obtaining the stability estimate with a weight that attains maximal values at the points where $v(x,t)$ is given (in this example, at $t=0$) and quickly decreases with increasing distance from the manifold where the Cauchy data are known.

Example 1.8. In some cases the ill-posed problem $Au = f$, $A \in \mathcal{L}(X)$, $\operatorname{Ker} A = \{0\}$, may be considered as the problem of finding the unbounded operator $B = A^{-1}$ at the element $f : u = A^{-1}f$. We shall now give a simple sufficient condition of l-well-posedness of this problem in terms of the resolvent $(\lambda - B)^{-1}$ of the operator B.

Suppose that for all $\lambda > 0$ the following inequality holds:

$$\|(\lambda - B)^{-1}\| \le m(\lambda), \qquad (1.24)$$

where

$$m'(\lambda) < 0, \qquad \lim_{\lambda\to\infty} m(\lambda) = 0. \qquad (1.25)$$

Then the estimate

$$\|Bf\| \le \varepsilon\|B^2 f\| + c(\varepsilon)\|f\| \qquad (1.26)$$

holds for all $f \in D(B^2)$ or, equivalently,

$$\|u\| \le \varepsilon\|A^{-1}u\| + c(\varepsilon)\|Au\|,$$

Chapter 5. Theory of conditionally well-posed problems

for all $u \in X \cap R(A)$, where $c(\varepsilon) = m^{-1}(\varepsilon) + m^{-2}(\varepsilon)\varepsilon$.

Thus, under the conditions (1.24) and (1.25) the problem $Au = f$ is l-well-posed and $l(u) = \|A^{-1}u\|$.

To prove estimate (1.26) it suffices to apply the triangle inequality to the identity

$$Bf = -(\lambda - B)^{-1}B^2 f - \lambda f + \lambda^2(\lambda - B)^{-1}f$$

and use conditions (1.24), (1.25) with $\varepsilon = m(\lambda)$ and $\lambda = m^{-1}(\varepsilon)$. In particular, if $X = C[0, T]$,

$$(Au)(t) = -\int_0^t u(\tau)\,d\tau = f(t)$$

then

$$Bf = -f'(t), \qquad f(0) = 0,$$
$$D(B) = R(A) = C^1[0, T] \cap \{f \mid f(0) = 0\},$$
$$(\lambda - B)^{-1}g = \int_0^t \exp(-\lambda(t - \tau))g(\tau)\,d\tau,$$
$$\|(\lambda - B)^{-1}\| \leq \lambda^{-1}, \qquad c(\varepsilon) = 2\varepsilon^{-1}.$$

Exercise 1.1. Suppose X and Y are Banach spaces and $A : D(A) \subseteq X \to R(A) \subseteq Y$ is a linear operator with domain of definition $D(A)$ and range of values $R(A)$. The linear operator A is said to be *closed* if $u_n \in D(A)$, $\|u_n - u\| \to 0$, $\|Au_n - f\| \to 0$ for $n \to \infty$ implies

$$u \in D(A), \qquad Au = f.$$

Prove that the problem $Au = f$ with closed operator is well-posed if and only if Ker $A = \{0\}$ and the range of values $R(A)$ is closed in Y, i.e., $R(A) = \overline{R(A)}$. In particular, if $\overline{R(A)} = Y$ then $R(A) = Y$. (Directions: use the closed graph theorem, which implies that any closed operator defined everywhere in a Banach space is bounded.)

Exercise 1.2. Assume that the domain of definition of a linear operator $A : D(A) \subset X \to Y$ is dense in the Banach space X. Suppose g is an element of the space Y^* conjugate to the Banach space Y. Then the linear functional $u \to \langle Au, g \rangle$ is defined in $D(A)$, where $\langle \cdot, \cdot \rangle$ is the duality between Y and Y^*. (In particular, if Y is a Hilbert space, then $\langle \cdot, \cdot \rangle$ is a scalar product in Y).

If this functional is bounded, then we say that g belongs to the domain of definition of the conjugate operator $A^* : D(A^*) \subseteq Y^* \to X^*$. Here we assume as the definition that $\langle A^*g, u \rangle = \langle g, Au \rangle$, $u \in D(A)$. In the left-hand side of this formula, $\langle \cdot, \cdot \rangle$ is the duality between X and X^*. The operator A^* is said to be *conjugate* to A. Since $D(A)$ is dense in X, the operator A^* is defined uniquely. Prove the following statements:

1) $\operatorname{Ker} A^* = R(A)^\perp$, where $R(A)^\perp = \{g \in Y^* \mid \langle g, f \rangle = 0, \ \forall f \in R(A)\}$.
2) $\overline{R(A)} = Y \Leftrightarrow \operatorname{Ker} A^* = \{0\}$.
3) The well-posed problem $Au = f$ has a solution if and only if $f \perp \operatorname{Ker} A^*$, i.e., $\langle g, f \rangle = 0$ for all $g \in \operatorname{Ker} A^*$.

(Directions: according to Exercise 1.1, $R(A) = \overline{R(A)}$. Therefore, it suffices to prove that $(\operatorname{Ker} A^*)^\perp = R(A)$.)

Exercise 1.3. Suppose X is an infinite-dimensional Banach space and A is completely continuous operator in X. Prove that the problem $Au = f$ is ill-posed.

Exercise 1.4. Suppose the operator A is given by the formula

$$(Au)(x) = \int_{\Omega_1} A(x, y) u(y) \, dy, \qquad x \in \Omega_2, \tag{1.27}$$

where Ω_1, Ω_2 are bounded open sets in $\mathbb{R}^n$ and $A(x, y) \in C^\infty(\overline{\Omega}_2 \times \overline{\Omega}_1)$. Show that the problem $Au = f$, $A : C(\overline{\Omega}_1) \to C^m(\overline{\Omega}_2)$ is ill-posed for any integer $m \geq 0$.

Exercise 1.5. Suppose that the kernel $A(x, y)$ of the integral operator defined by formula (1.27), where $\Omega_1 = \Omega_2 = \Omega$, satisfies the estimate

$$|A(x, y)| \leq c|x - y|^{-\alpha}, \qquad 0 < \alpha < n.$$

Show that $A \in \mathcal{L}(L_2(\Omega))$ and the problem $Au = f$ is ill-posed. (Directions: prove that the operator A is completely continuous in $L_2(\Omega)$.)

Exercise 1.6. Let an operator A be defined by the formula

$$(Au)(t) = \int_0^t A(t, \tau) u(\tau) \, d\tau = f(t), \qquad t \in [0, T],$$

$$A(t, \tau) \in C_1(\overline{\Omega}), \qquad \Omega = \{t, \tau \in \mathbb{R}^1 \mid 0 < \tau < t < T\}.$$

Prove that the problem $Au = f$ is ill-posed in $C[0,T]$, but well-posed from $C[0,T]$ to $C^1[0,T]$ if we assume that $A(t,t) \neq 0$ for all $t \in [0,T]$.

Exercise 1.7. Suppose an operator $A_\alpha : L_1(\Omega_1) \to L_1(\Omega_2)$ is given by the formula

$$(A_\alpha u)(x) = \int_{\Omega_1} |x-y|^{\alpha-n} u(y)\, dy, \qquad x \in \Omega_2,$$

where Ω_1, Ω_2 are bounded sets in $\mathbb{R}^n$, $\Omega_1 \cap \Omega_2 = \emptyset$, $\alpha \in \mathbb{R}^1$. Prove that
1) the problem $A_\alpha u = f$ is ill-posed;
2) if $\alpha \neq 2+2m$, $\alpha \neq n+2m$ for all $m = 0, 1, 2, \ldots$, then $\operatorname{Ker} A_\alpha = \{0\}$.
(Directions: using the formula $\Delta |x|^{\alpha-n} = (\alpha-2)(\alpha-n)|x|^{\alpha-n-2}$, where $\Delta = \sum_{j=1}^{n} \partial_j^2$, prove by induction that if $A_\alpha u = 0$, then $\int_{\Omega_1} y^\beta u(y)\, dy = 0$ for any multi-index $\beta = (\beta_1, \beta_2, \ldots, \beta_n)$.)

Exercise 1.8. Establish ill-posedness of the following Cauchy problems in the class of functions of finite smoothness
1) $u_t = u_{xx}$, $x, t \in \mathbb{R}$, $x \geq 0$, $u(0,t) = g(t)$, $u_x(0,t) = 0$;
2) $u_{xx} + u_{yy} = 0$, $x, y \in \mathbb{R}$, $x \geq 0$, $u(0,y) = g(y)$, $u_x(0,y) = 0$;
3) $u_{tt} = u_{xx} + u_{yy}$, $x, y, t \in \mathbb{R}$, $x \geq 0$, $u(0,y,t) = g(y,t)$, $u_x(0,y,t) = 0$.

Exercise 1.9. Suppose X is a complex Banach space, $A \in \mathcal{L}(X)$ and $\operatorname{Ker} A = \{0\}$. Assume that along a certain ray $\lambda = r\exp(i\varphi)$, $r \geq r_0 \geq 0$, $\varphi = \text{const}$ the following estimate holds:

$$\|A(E - \lambda A)^{-1} u\| \leq m(r)\|u\|, \qquad \forall u \in X, \qquad (1.28)$$

where $m'(r) < 0$, $m(r) \to 0$ for $r \to \infty$.
Show that the problem $Au = f$ is l-well-posed and $l(u) = \|A^{-1}u\|$. Express the number $c(\varepsilon)$ in terms of the function $m(r)$.

Exercise 1.10. Suppose $X = C[0,T]$ and

$$(Au)(t) = \int_0^t u(\tau)\, d\tau, \qquad t \in [0,T].$$

For the operator $A \in \mathcal{L}(X)$, find the ray $\lambda = r\exp(i\varphi)$ along which estimate (1.28) holds with $m(r) = r^{-1}$.

Exercise 1.11. Prove that if the problem $Au = f$ is l-well-posed, i.e., estimate (1.11) holds, then we may assume that $c(\varepsilon) \in C^\infty(0, \infty)$ without loss of generality.

Exercise 1.12. For an l-well-posed problem $Au = f$, we set

$$\omega_m(\delta) = \inf_{\varepsilon > 0}(\varepsilon m + c(\varepsilon)\delta).$$

Show that
1) if $c(\varepsilon) = c_0 \varepsilon^{-\nu}$, $\nu\, 0$, then

$$\omega_m(\delta) \sim c_1 \delta^{1/(1+\nu)} m^{\nu/(1+\nu)}, \qquad \delta \to 0,$$

$$c_1 = \nu^{-1}(1+\nu)(c_0\nu)^{1/(1+\nu)};$$

2) if $c(\varepsilon) = c_0 \exp(s\varepsilon^{-\nu})$, $\nu > 0$, then

$$\omega_m(\delta) \sim m(s/\ln\delta^{-1})^{1/\nu}, \qquad \delta \to 0.$$

Exercise 1.13. The problem $Au = f$ is called *Tikhonov-well-posed in the set* $M_m = \{u \in D(l) \mid l(u) \leq m\}$ if $\|u\| \leq \omega_m(\|Au\|)$ for all $u \in M$, where the function $\omega_m(\delta)$ of the parameter $\delta \geq 0$ is nonincreasing and continuous at the zero point, and $\omega_m(0) = 0$. Prove that the problem $Au = f$ is l-well-posed if and only if it is Tikhonov-well-posed in any set M_m, $m > 0$.

5.2. l_h-WELL-POSEDNESS OF DIFFERENCE SCHEMES

For numerical solution of equation (1.2) by finite-difference methods, normed spaces X_h, Y_h of mesh functions depending on the vector h are introduced and an operator $A_h \in \mathcal{L}(X_h, Y_h)$ which approximates the operator A in a certain sense is constructed. As a result, we obtain the difference equation

$$A_h u_h = f_h. \tag{2.1}$$

The operator equation (2.1), which depends on the parameter h, is called a *difference scheme*, following the terminology of Samarskii (1977).

Definition 2.1. The difference scheme $A_h u_h = f_h$ is called *well-posed* if

$$R(A_h) = Y_h$$

Chapter 5. Theory of conditionally well-posed problems

for sufficiently small $|h| \leq h_0$ and the estimate

$$\|u\| \leq c\|A_h u\|, \tag{2.2}$$

holds for all $u \in X_h$, where the number c is independent of h, $|h| \leq h_0$. (We do not specify in what spaces the norms are taken if it is clear from the context.)

From this definition it follows that the difference scheme is well-posed if and only if for all sufficiently small h there exists an inverse operator $A_h^{-1} \in \mathcal{L}(Y_h, X_h)$ whose norm is uniformly bounded with respect to h:

$$\|A_h^{-1}\| \leq c.$$

Note that, in contrast to Definition 1.1, we require here that the range of values of the operator A_h should coincide with the whole space Y_h. Thus, the existence theorem is postulated. The difference schemes satisfying only condition (2.2) are called *stable*. It is intuitively evident that for any reasonable definition of approximation the ill-posed problem (1.2) cannot be approximated by a stable difference scheme. Therefore, in the case where $Au = f$ is only l-well-posed it is natural to introduce a weaker definition of stability of a difference scheme, which is the difference analogue of estimate (1.11). So we assume that, in addition to the space X_h, we have the linear space $D(l_h)$ with some seminorm l_h such that $X_h \cap D(l_h) \neq \emptyset$. In applications we usually have $D(l_h) \subseteq X_h$ or, conversely, the space X_h can be identified with a certain subspace in $D(l_h)$.

Definition 2.2. The difference scheme (2.1) is called l_h-*well-posed* if

$$R(A_h) = Y_k,$$

and for any $\varepsilon \to 0$ there exists a number $c(\varepsilon) > 0$ independent of h such that for any $u \in X_h \cap D(l_h)$ and sufficiently small h, $|h| \leq h_0$, the following estimate holds:

$$\|u\| \leq \varepsilon l_h(u) + c(\varepsilon)\|A_h u\| \tag{2.3}$$

Difference schemes that satisfy estimate (2.3) are called l_h-*stable* and the functional l_h is called *the stabilizing functional*.

Thus, the definition of l_h-stability is a generalization of the classical stability condition (2.2) for ill-posed problems.

We now introduce the notations necessary for definition of approximation and convergence. For this purpose, we assume that there exist linear operators

$$P_h^1 : X \to X_h, \qquad P_h^2 : Y \to Y_h$$

such that
$$\lim_{h\to 0} \|P_h^1 u\| = \|u\|_X, \quad \forall u \in X, \tag{2.4}$$

$$\lim_{h\to 0} \|P_h^2 u\| = \|f\|_Y, \quad \forall f \in Y. \tag{2.5}$$

The operators P_h^j in specific examples are the operators of projection on the mesh. We say that a sequence $u_h \in X_h$ converges to the element $u \in X$ if
$$\lim_{h\to 0} \|P_h^1 u - u_h\| = 0.$$

If
$$\|P_h^1 u - u_h\| \le c\psi(h),$$
for sufficiently small $|h|$, where $\lim_{|h|\to 0} \psi(h) = 0$ and the number c is independent of h, we say that the sequence $u_h \in X_h$ converges to $u \in X$ with rate $O(\psi(h))$. The analogous notion of convergence of $f_h \in Y_h$ to the element $f \in Y$ means that
$$\lim_{|h|\to 0} \|P_h^2 f - f_h\| = 0.$$

The operator A_h approximates the operator A in the set $V \subseteq D(A)$ if
$$\lim_{|h|\to 0} \|A_h P_h^1 u - P_h^2 A u\| = 0$$
for any element $u \in V$.

If for any $u \in V$ and sufficiently small h, $|h| \le h_0$, we have
$$\|A_h P_h^1 u - P_h^2 A u\| \le c\psi(h),$$
where the number c is independent of h, then we say that the operator A_h has the approximation order $O(\psi(h))$ in the set V.

Now, we consider the problem of approximate solution of equation (1.2)
$$Au = f$$
with approximate right-hand side f_δ:
$$\|P_n^2(f + f_\delta)\| \le \delta. \tag{2.6}$$

We solve this problem using the l_h-well-posed difference scheme
$$A_h u_{h\delta} = f_{h\delta}, \quad f_{h\delta} = P_h^2 f_\delta. \tag{2.7}$$

Chapter 5. Theory of conditionally well-posed problems 149

By the triangle inequality, we have

$$\begin{aligned}\|u_{h\delta} - P_h^1 u\| &= \|A_h^{-1}(P_h^2 f_\delta - P_h^2 f + P_h^2 f) - P_h^1 u\| \\ &\leq \|A_h^{-1}\| \|P_h^2(f_\delta - f)\| + \|A_h^{-1} P_h^2 f - P_h^1 u\| \\ &\leq \|A_h^{-1}\|\delta + \|A_h^{-1}(P_h^2 A u - A_h P_h^1 u)\| \\ &\leq \|A_h^{-1}\|\delta + \varepsilon l_h(A_h^{-1} P_h^2 A u - P_h^1 u) + c(\varepsilon)\|(P_h^2 A - A_h P_h^1 u)\|.\end{aligned} \quad (2.8)$$

This estimate yields the following theorem.

Theorem 2.1. *Suppose that we have an l_h-well-posed difference scheme and the operator A_h approximates the operator A on the solution of equation (1.2) with order $\psi(h)$, i.e.,*

$$\|(A_h P_h^1 - P_h^2 A)u\| \leq \psi(h), \qquad |h| \leq h_0 \quad (2.9)$$

and, moreover,

$$l_h(A_h^{-1} P_h^2 A u - P_h^1 u) \leq m(h) \leq m_1, \quad (2.10)$$

where the number m_1 is independent of h. Further, suppose that the condition of coordination of the parameter h with the error δ of initial data holds

$$\|A_h^{-1}\|\delta \to 0. \quad (2.11)$$

Then the solution $u_{h\delta}$ of equation (2.7) converges to the solution u of the equation $Au = f$ as $|h| \to 0$, and

$$\|u_{h\delta} - P_h^1 u\| \leq \|A_h^{-1}\|\delta + \omega_m(\psi(h)).$$

Here

$$\omega_m(\alpha) = \inf_{\varepsilon > 1}(\varepsilon m + c(\varepsilon)\alpha), \quad (2.12)$$

$$\omega_m(\alpha) \leq \omega_{m_1}(\alpha) \to 0 \quad \text{for} \quad \alpha \to 0.$$

To prove this, it suffices to take the infimum with respect to ε of both sides of inequality (2.8) and take into account conditions (2.9)–(2.11).

Remark 2.1. Condition (2.10) will hold if, for example,

$$l_h(P_h^1 u) \leq m/2, \quad (2.13)$$

$$l_h(A_h^{-1} P_h^2 A u) \leq m/2. \quad (2.14)$$

In some problems, condition (2.14) follows from (2.13). Conditions (2.13), (2.14), and (2.10) can be called *the conditions of a priori smoothness of a solution*.

The theorem stating that the properties of stability and approximation guarantee the convergence of the difference scheme is evidently generalized to the ill-posed case by Theorem 2.1. More general condition of convergence, which does not involve the notion of l_h-stability, is given by the following statement.

Proposition 2.1. Suppose $A_h^{-1} \in \mathcal{L}(Y_h, X_h)$, the parameters h and δ are coordinated so that

$$\|A_h^{-1}\|\delta \to 0 \quad \text{for} \quad \delta \to 0, \ |h| \to 0,$$

and for a solution u of equation (1.2) we have

$$\|A_h^{-1} P_h^2 Au - P_h^1 u\| \to 0 \quad \text{for} \quad |h| \to 0. \tag{2.15}$$

Then

$$\|u_{h\delta} - P_h^1 u\| \leq \|A_h^{-1}\|\delta + \|A_h^{-1} P_h^2 Au - P_h^1 u\| \to 0.$$

Note that condition (2.15) implicitly implies the uniqueness of the solution of the equation $Au = f$. Otherwise, there exists an element $u \in D(A)$, $u \neq 0$ such that $Au = 0$, and, consequently, $A_h^{-1} P_h^2 Au - P_h^1 u = -P_h^1 u$, $\|P_h^1 u\| \to \|u\| \neq 0$. In contrast, in Theorem 2.1 the *a priori* uniqueness of solution of the equation $Au = f$ is not assumed. It holds as a corollary of the condition of l_h-stability and approximation. Indeed, let

$$D_m(A) = \left\{ u \in D(A) \mid P_h^1 u \in D(l_h), \ l_h(P_h^1 u) \leq m \right\},$$

where the number m is independent of h for $|h| \leq h_0$. Set

$$D_\infty(A) = \bigcup_{m=1}^{\infty} D_m(A),$$

$$l(u) = \sup_{|h| \leq h_0} l_h(P_h^1 u), \quad u \in D_\infty(A) = D(l).$$

Then from the l_h-stability condition and approximation (2.9) we obtain

$$\|P_h^1 u\| \leq \varepsilon l_h(P_h^1 u) + c(\varepsilon)\|A_h P_h^1 u\|$$
$$\leq \varepsilon l(u) + c(\varepsilon)[\|P_h^2 Au\| + \|A_h P_h^1 u - P_h^2 Au\|]$$
$$\leq \varepsilon l(u) + c(\varepsilon)[\|P_h^2 Au\| + \psi(h)].$$

Chapter 5. Theory of conditionally well-posed problems

Passing to the limit as $|h| \to 0$ and taking into account (2.4) and (2.5), we get
$$\|u\|_X \le \varepsilon l(u) + c(\varepsilon)\|Au\|_Y,$$
i.e., the problem $Au = f$ is l-well-posed. In particular, the solution of this equation for $u \in D(l)$ is unique. Thus, the theory of l_h-stable difference schemes not only provides the algorithms of approximate solution of ill-posed problems, but also allows proving the uniqueness theorems and the estimates of conditional stability of the original problem in the exact statement. The systematic description of the sufficient conditions for the l_h-stability of difference schemes for many kinds of ill-posed boundary value problems will be given in the next chapter. Here we shall give only an example of the problem of numerical differentiation.

Example 2.1. Suppose that a normed space X consists of functions
$$u(t) \in C[0,T] \cap L_2(0,T), \qquad \|u\|_X = \int_0^T |u(t)|^2 \, dt.$$

We define an operator $A \in \mathcal{L}(X)$ by the formula
$$(Au)(t) = \int_0^t u(\tau) \, d\tau, \qquad t \in [0,T].$$

Set $P_h^1 = P_h^2 = P_h$,
$$(P_h u)_j = u(jh), \qquad j = 1, \ldots, N, \quad Nh = T, \tag{2.16}$$

$$(A_h v)_j = h \sum_{k=1}^j v_k, \tag{2.17}$$

where
$$v = \begin{pmatrix} v_1 \\ v_2 \\ \vdots \\ v_N \end{pmatrix} \in l_{2,h}(1,N) = X_h,$$

$$\|v\|_{l_{2,h}(1,N)}^2 = h \sum_{j=1}^N |v_j|^2 = \|v\|^2.$$

Define a scalar product in the space $l_{2,h}(1, N)$ by the formula

$$\langle v, w \rangle = h \sum_{j=1}^{N} v_j w_j, \quad v, w \in X_h.$$

Following (2.17), the operator A_h is represented in the form

$$A_h = h \begin{pmatrix} 1 & 0 & 0 & . & 0 & 0 \\ 1 & 1 & 0 & . & 0 & 0 \\ . & . & . & . & . & . \\ 1 & 1 & 1 & . & 1 & 0 \\ 1 & 1 & 1 & . & 1 & 1 \end{pmatrix}.$$

Hence, it is easy to see that

$$(A_h^{-1})^* = \frac{1}{h} \begin{pmatrix} 1 & -1 & 0 & . & 0 & 0 \\ 0 & 1 & -1 & . & 0 & 0 \\ . & . & . & . & . & . \\ 0 & 0 & 0 & . & 1 & -1 \\ 0 & 0 & 0 & . & 0 & 1 \end{pmatrix}. \quad (2.18)$$

According to the general scheme of Example 1.6 (see (1.19)), we have

$$\|v\| \leq \varepsilon l_h(v) + (2\varepsilon)^{-1} \|A_h v\|, \quad \forall v \in X_h,$$

where

$$l_h^2(v) = \|(A_h^{-1})^* v\|^2 = h \sum_{j=1}^{N-1} |(v_{j+1} - v_j)/h|^2 + v_N^2/h.$$

Besides, it is clear that $A_h X_h = X_h$. Thus, the difference scheme defined by formula (2.17) is l_h-well-posed. Estimate the error of approximation, $\|(P_h A - A_h P_h)u\|$. By definition,

$$(P_h A u)_j = \int_0^{jh} u(\tau)\,d\tau, \quad (A_h P_h u)_j = h \sum_{k=1}^{j} u(kh),$$

which implies that

$$(P_h A u - A_h P_h u)_j = \sum_{k=1}^{j} \left(\int_{(k-1)h}^{kh} u(\tau)\,d\tau - h u(kh) \right)$$

$$= \sum_{k=1}^{j} \int_{(k-1)h}^{kh} (u(\tau) - u(kh))\,d\tau.$$

Chapter 5. Theory of conditionally well-posed problems

For $\|u'\|_{C[0,T]} \leq M$ we have $|u(\tau) - u(kh)| \leq M|\tau - kh|$. Therefore,

$$|(P_h Au - A_h P_h u)_j| \leq M \sum_{k=1}^{N} h^2/2 = MhT/2.$$

$$\|P_h Au - A_h P_h u\| \leq \left\{ h \sum_{j=1}^{N} (MhT/2)^2 \right\}^{1/2} \leq MhT^{3/2}/2.$$

Similarly,

$$(A_h^{-1} P_h Au - P_h u)_j = \frac{1}{h} \int_{(j-1)h}^{jh} u(\tau)\, d\tau - u(ih) = \frac{1}{h} \int_{(j-1)h}^{jh} (u(\tau) - u(jh))\, d\tau,$$

$$|(A_h^{-1} P_h Au - P_h u)_j| \leq Mh/2, \qquad \|A_h^{-1} P_h Au - P_h u\| \leq Mh\sqrt{T}/2.$$

From (2.18) it follows that

$$\|u_{h\delta} - P_h u\| \leq \|A_h^{-1}\|\delta + \|A_h^{-1} P_h Au - P_h u\| \leq 2\delta/h + Mh\sqrt{T}/2. \quad (2.19)$$

This estimate shows that the optimal step for the given noise level δ will be $h = 2(\delta/MT^{1/2})^{1/2}$, which is obtained by minimization of the right-hand side of inequality (2.19) with respect to h. For such choice of h, we have

$$\|u_{h\delta} - P_h u\| \leq 2(M\delta T^{1/2})^{1/2}.$$

Exercise 2.1. Suppose that the functional l_h satisfies the estimate

$$l_h(v) \leq \mathcal{L}_h \|A_h v\|, \qquad \forall v \in X_h$$

and conditions (2.9) and (2.10) hold. Also, we suppose that the parameters h and δ are coordinated as follows:

$$\mathcal{L}_h \delta \leq m_2,$$

where m_2 is independent of h and δ. Show that the solution of equation (2.7) converges to the solution u of equation (1.2) and

$$\|u_{h\delta} - P_h' u\| \leq \omega_{m_1+m_2}(\delta + \psi(h)),$$

where the function ω_m is defined by formula (2.12).

Exercise 2.2. Verify condition (2.4) for the operator P_h defined by formula (2.16). How should we change definition (2.16) of the operator P_h in order for condition (2.4) to hold for $X = L_2(0,T)$?

Exercise 2.3. In Example 2.1, suppose $X = C[0,T]$. Set

$$\|v\|_{X_h} = \max_{j=1,\ldots,N} |v_j|, \qquad v \in \mathbb{R}^N, \quad hN = T,$$

$$(B_h v)_j = -(v_j - v_{j-1})/h, \qquad v_0 = 0, \quad j = 1, \ldots, N.$$

In other words, $B_h = -A_h^{-1}$, where the operator A_h is defined by (2.17). Prove that

$$\|(\lambda - B_h)^{-1}\| \leq \lambda^{-1}, \qquad B_h \in \mathcal{L}(X_h). \tag{2.20}$$

for $\lambda > 0$. Using the method of Example 1.7, show that

$$\|B_h v\| \leq \varepsilon \|B_h^2 v\| + 2\varepsilon^{-1} \|v\|, \qquad \|v\| \leq \varepsilon \|A_h^{-1} v\| + 2\varepsilon^{-1} \|A_h v\|.$$

Exercise 2.4. Suppose that the spaces X, X_h, and the operator P_h are the same as in Exercise 2.3. Define an operator B_h by the formula

$$(B_h v)_j = -(v_{j+1} - v_{j-1})/2h, \qquad v_0 = 0, \quad v_{N+1} = 0.$$

What is the order of approximation of the operator B by the operator B_h in the set of functions $C^3[0,T]$. For which $\lambda > 0$ the estimate (2.20) can be guaranteed?

5.3. VARIATIONAL METHODS OF SOLUTION OF l_h-STABLE DIFFERENCE SCHEMES

Suppose v is an arbitrary element of the space X_h introduced in Section 2. Using the definition of l_h-stability of a difference scheme with operator A_h approximating A on the solution u, we shall try to find an element v such that

$$\|v - P_h^1 u\| \to 0 \quad \text{for} \quad |h| \to 0$$

by the given element $f_{h\delta} = P_h^2 f_\delta$. Estimates (2.3), (2.6), and (2.9) yield

$$\|v - P_h^1 u\| \leq \varepsilon l_h(P_h^1 u) + \varepsilon l_h(v)$$
$$+ c(\varepsilon) \|A_h v - A_h P_h^1 u + P_h^2 A u - P_h^2 A u + P_h^2 f_\delta - P_h^2 f_\delta\|$$
$$\leq \varepsilon l_h(P_h^1 u) + \varepsilon l_h(v) + c(\varepsilon)[\|A_h v - f_{h\delta}\| + \psi(h) + \delta]. \tag{3.1}$$

Chapter 5. Theory of conditionally well-posed problems

If the difference scheme is well-posed, i.e., $l_h = 0$ and $c(\varepsilon) = \text{const}$, then the best element v (from the viewpoint of minimization of the right-hand side of inequality (3.1)) will be the solution of the equation $A_h v = f_{h\delta}$, i.e., $v = A_h^{-1} f_{h\delta}$. In this case,

$$\|v - P_h^1 u\| \le c(\delta + \psi(h)).$$

In the general case, it is natural to minimize with respect to v the functional

$$\Phi_1(v,\varepsilon) = \varepsilon l_h(v) + c(\varepsilon)\|A_h v - f_{h\delta}\|. \tag{3.2}$$

But in practice it is easier to minimize the quadratic functional

$$\Phi_2(v,\varepsilon) = \varepsilon^2 l_h^2(v) + c^2(\varepsilon)\|A_h v - f_{h\delta}\|^2$$

or the functional

$$J(v,\alpha) = \alpha l_h^2(v) + \|A_h v - f_{h\delta}\|^2, \tag{3.3}$$

where

$$\alpha = \varepsilon^2 / c^2(\varepsilon). \tag{3.4}$$

The parameters h, δ, and α are chosen such that the expression in the right-hand side of inequality (3.1) tends to zero. Since

$$\Phi_2(v,\varepsilon) \le \Phi_1^2(v,\varepsilon) \le 2\Phi_2(v,\varepsilon), \tag{3.5}$$

passing from minimization of the functional $\Phi_1(v,\varepsilon)$ to $\Phi_2(v,\varepsilon)$ does not influence estimate (3.1) considerably. So, we suppose that $v_{\delta h}$ is the element which provides the minimum with respect to v of the functional $\Phi_2(v,\varepsilon)$ or, what is the same, $J(v,\alpha)$:

$$\Phi_2(v_{h\delta},\varepsilon) \le \Phi_2(v,\varepsilon), \qquad \forall v \in X_h \cap D(l_h).$$

Then, in particular, $\Phi_2(v_{h\delta},\varepsilon) \le \Phi_2(P_h^1 u,\varepsilon)$. Therefore (see (3.5)), we have

$$\Phi_1(v_{h\delta},\varepsilon) \le \sqrt{2}\Phi_2^{1/2}(P_h^1 u,\varepsilon) \le \sqrt{2}\Phi_1(P_h^1 u,\varepsilon). \tag{3.6}$$

On the other hand, (3.2), (2.6), and (2.9) yield

$$\Phi_1(P_h^1 u,\varepsilon) = \varepsilon l_h(P_h^1 u) + c(\varepsilon)\|A_h P_h^1 u - f_{h\delta}\|$$
$$= \varepsilon l_h(P_h^1 u) + c(\varepsilon)\|A_h P_h^1 u - P_h^2 A u + P_h^2 f - P_h^2 f_\delta\|$$
$$\le \varepsilon l_h(P_h^1 u) + c(\varepsilon)[\psi(h) + \delta]. \tag{3.7}$$

As a result, taking into account (3.6), (3.7), and (3.1), we obtain

$$\|v_{h\delta} - P_h^1 u\| \leq \{\varepsilon l_h(P_h^1 u) + c(\varepsilon)[\psi(h) + \delta]\}(1 + \sqrt{2}). \tag{3.8}$$

Let

$$l_h(P_h^1 u) \leq m, \tag{3.9}$$

where the number m is independent of h, and

$$\min_{\varepsilon \geq 0} F(\varepsilon) = F(\varepsilon_0) = \omega_m(\psi(h) + \delta), \tag{3.10}$$

$$F(\varepsilon) = \varepsilon m + c(\varepsilon)(\psi(h) + \delta). \tag{3.11}$$

Then (3.8) implies the estimate

$$\|v_{h\delta} - P_h^1 u\| \leq \omega_m(\psi(h) + \delta)(1 + \sqrt{2}) \tag{3.12}$$

Thus, we have proved the following theorem.

Theorem 3.1. *Suppose the difference scheme is l_h-stable and the approximation conditions (2.9) and the conditions (3.9) of the a priori smoothness of solution hold. If $v_{h\delta}$ is a solution of the extremal problem*

$$J(v_{h\delta}, \alpha) = \min_v J(v, \alpha), \qquad v \in X_h \cap D(l_h),$$

where the parameter $\alpha = \alpha(\varepsilon_0) = \varepsilon_0^2/c^2(\varepsilon_0)$ is determined from conditions (3.10) and (3.11), then the sequence $v_{h\delta}$ converges to the solution of the equation $Au = f$ as $|h| \to 0$, $\delta \to 0$. Moreover, estimate (3.12) holds.

The functional $J(v, \alpha)$ defined by formula (3.3) is often called *the smoothing functional*. The parameter $\alpha > 0$ in this functional is called *the regularization parameter*. In Theorem 3.1 this parameter is defined by the formula $\alpha = \alpha(\varepsilon_0)$, in which it is assumed that the functions $\psi(h)$, $c(\varepsilon)$ and the numbers δ and m are known. In the following theorem, it is not necessary to know the function $c(\varepsilon)$ and the number m when choosing α.

Theorem 3.2. *Suppose $v_{h\delta}$ is the element for which the minimum of the smoothing functional $J(v, \alpha)$ is achieved:*

$$J(v_{h\delta}, \alpha) \leq J(v, \alpha), \qquad \forall v \in X_h \cap D(l_h).$$

If the condition $(\delta + \psi(h))/\sqrt{\alpha} = O(1)$ of coordination of the parameters α, δ, and h holds and, moreover, $l_h(P_h^1 u) = O(1)$, then $\|v_{h\delta} - P_h^1 u\| \to 0$ as $\delta, \alpha, h \to 0$.

Chapter 5. Theory of conditionally well-posed problems   157

Proof. From estimate (3.8), taking into account (3.4), we obtain

$$\|v_{h\delta} - P_h^1 u\| \leq (1 + \sqrt{2})\varepsilon[l_h(P_h^1 u) + (\delta + \psi(h))/\sqrt{\alpha}].$$

As in Section 1, without loss of generality, we may assume that $c(\varepsilon)$ is a continuous nonincreasing function depending on ε such that $c(0) = \infty$ and $c(\infty) = \text{const}$. In this case, there exists a unique root $\varepsilon = \varepsilon(\alpha)$ of equation (3.4):

$$g(\varepsilon(\alpha)) = \varepsilon - c(\varepsilon)\sqrt{\alpha} = 0.$$

Evidently, $\varepsilon(\alpha) \to 0$ as $\alpha \to 0$. Thus,

$$\|v_{h\delta} - P_h^1 u\| = O(\varepsilon) \to 0 \quad \text{for} \quad \alpha \to 0.$$

□

In applications, the stabilizing functional $l_h(v)$ characterizes the smoothness of the function v. The smaller is the functional, the smoother is the function v. In particular, if we need to find the smoothest solution of the equation $Au = f$ within the limits of accuracy δ and $\psi(h)$, then we may consider the solution of the following variational problem

$$\min l_h(v)$$

under the condition

$$\|A_h v - f_{h\delta}\| \leq \delta + \psi(h). \tag{3.13}$$

Suppose $v_{h\delta}$ is a solution of this extremal problem. Then, by definition, $l_h(v_{h\delta}) \leq l_h(v)$ for all v that belong to the set defined by inequality (3.13). Since

$$\|A_h P_h^1 u - f_{h\delta}\| = \|A_h P_h^1 u - P_h^2 A u + P_h^2 f - P_h^2 f_\delta\| \leq \delta + \psi(h),$$

we have

$$l_h(v_{h\delta}) \leq l_h(P_h^1 u),$$

and estimate (3.1) yields

$$\|v_{h\delta} - P_h^1 u\| \leq 2[\varepsilon l_h(P_h u) + c(\varepsilon)(\psi(h) + \delta)].$$

Taking the infimum of the right-hand side with respect to ε, we obtain

$$\|v_{h\delta} - P_h^1 u\| \leq 2\omega_m(\delta + \psi(h)) \to 0, \qquad \delta \to 0, \quad h \to 0,$$

under the condition that $l_h(P_h^1 u) \leq m$.

Note that, using the Lagrange method of finding the conditional extremum under sufficiently general assumptions for l_h and A_h, we can show that $v_{h\delta}$ is determined from the condition for the minimum of the smoothing functional $J(v, \alpha)$:

$$J(v_\alpha, \alpha) = \min J(v, \alpha), \qquad \forall v \in X_h \cap D(l_h), \tag{3.14}$$

where the parameter α is found from the condition

$$\|A_h v_\alpha - f_{h\delta}\| = \delta + \psi(h). \tag{3.15}$$

Example 3.1. We find the element v_α in explicit form from condition (3.14) in the case where $X_h = Y_h$ and X_h is a real Hilbert space with scalar product $\langle \cdot, \cdot \rangle$ and

$$l_h(v) = \|L_h v\| \geq \mu \|v\|, \qquad \mu > 0, \quad L_h \in \mathcal{L}(X_h). \tag{3.16}$$

By the definition of v_α, we have

$$J(v_\alpha + tw, \ \alpha) \geq J(v_\alpha, \alpha)$$

for all real t and any $w \in X_h$.

Thus, for a fixed w the real function $\varphi(t) = J(v_\alpha + tw, \alpha)$ reaches its minimum at $t = 0$. Therefore, $\varphi'(0) = 0$. Since

$$\begin{aligned}J(v_\alpha + tw, \alpha) &= \alpha \langle L_h v_\alpha + t L_h w, L_h v_\alpha + t L_h w \rangle \\ &\quad + \langle A_h v_\alpha + t A_h w - f_{h\delta}, A_h v_\alpha + t A_h w - f_{h\delta} \rangle \\ &= a_0 + a_1 t + a_2 t^2\end{aligned}$$

is a polynomial of degree 2 with respect to t, then $\varphi'(0) = 0$ is equivalent to $a_1 = 0$. However,

$$\begin{aligned}a_1 &= 2\alpha \langle L_h v_\alpha, L_h w \rangle + 2 \langle A_h v_\alpha - f_{h\delta}, A_h w \rangle \\ &= 2 \langle \alpha L_h^* L_h v_\alpha + A_h^* A_h v_\alpha - A_h^* f_{h\delta}, w \rangle = 0.\end{aligned}$$

Since the element w is arbitrary, we obtain the Euler equation for v_α:

$$(\alpha L_h^* L_h + A_h^* A_h) v_\alpha = A_h^* f_{h\delta}. \tag{3.17}$$

Since $a_2 = \alpha \|L_h w\|^2 + \|A_h w\|^2 \geq \alpha \mu^2 \|w\|^2 > 0$ for $w \neq 0$, the minimum is reached, indeed, at the element

$$v_\alpha = (\alpha L_h^* L_h + A_h^* A_h)^{-1} A_h^* f_{h\delta},$$

Chapter 5. Theory of conditionally well-posed problems

and there are no other extremal elements. By condition (3.16), we have

$$(\alpha L_h^* L_h + A_h^* A_h)^{-1} \in \mathcal{L}(X_h).$$

Note that, when proving the convergence of variational methods of solution of the ill-posed problem $Au = f$, we have used the l_h-stability condition for the difference scheme $A_h u_h = f_{h\delta}$. The fact that $R(A_h) = Y_h$ has not been used.

Exercise 3.1. Write the Euler equation (3.17) for the problem of differentiation from Example 2.1.

Exercise 3.2. Suppose $l_h(P_h^1 u) \leq m$. The squared estimate (3.1) leads to the inequality

$$\|v - P_h^1 u\|^2 \leq 4\{\varepsilon^2 m^2 + \varepsilon^2 l_h^2(v) + c^2(\varepsilon)\|A_h v - f_{h\delta}\|^2 + c^2(\varepsilon)(\psi(h) + \delta)^2\}.$$

We denote the functional in braces by $J(v, \varepsilon)$. Taking into account the estimate $\|v - P_h^1 u\|^2 \leq 4 J(v, \varepsilon)$, it is natural to determine the element v from the condition of minimum of the functional $J(v, \varepsilon)$ with respect to $v \in X_h \cap D(l_h)$ and $\varepsilon > 0$. Prove that if

$$l_h(v) = \|L_h v\|_{X_h} \geq \mu \|v\|_{X_h}, \qquad L_h \in \mathcal{L}(X_h),$$

and X_h is a real Hilbert space, then the element v and parameter ε which minimize the functional $J(v, \varepsilon)$ are determined from the following system of equations:

$$(\alpha L_h^* L_h + A_h^* A_h)v = A_h^* f_{h\delta}, \qquad (3.18)$$

$$\alpha = \varepsilon^2/c^2(\varepsilon), \qquad (3.19)$$

$$\varepsilon^{-1}\alpha(m^2 + \|L_h v\|^2) = -(c(\varepsilon))^{-1} c'(\varepsilon)\Big[\|A_h v - f_{h\delta}\|^2 + (\delta + \psi(h))^2\Big]. \qquad (3.20)$$

The function $c(\varepsilon)$ entering the l_h-stability estimate is assumed to be differentiable. In particular, for $c(\varepsilon) = c_0 \varepsilon^{-\beta}$, $\beta > 0$, the last equation takes the form

$$\alpha(m^2 + \|L_h v\|^2) = \beta[\|A_h v - f_{h\delta}\|^2 + (\delta + \psi(h))].$$

Exercise 3.3. Substantiate the method of construction of the approximate solution v presented in Exercise 3.2. In other words, find the conditions that guarantee the existence of a unique solution of the system (3.18)–(3.20) and show that $\|v - P_h^1 u\| \to 0$ as $h \to 0$, $\delta \to 0$.

5.4. REMARKS AND REFERENCES

Basic results of this chapter are well known. However, their systematic presentation on the basis of the notion of l-well-posedness (Bukhgeim, 1972) is a new feature. This definition is dual, in a certain sense, to the definition of the Tikhonov well-posedness introduced by Lavrent'ev (1962) (see Exercise 1.13). In our definition, as is easy to see, the functions $\omega_m(\delta)$ and $-c^{-1}(p)$, where $p = c(\varepsilon)$, are Young-dual (i.e., connected by the Legendre transform). The definition is correct both in the case where additional information on solution has quantitative character $l(u) \leq m$ and in the case of just qualitative information on the smoothness of the solution $u \in D(l)$.

To Section 1. For the questions concerning Definition 1.1, Exercises 1.1 and 1.2, see, for example, Krein (1971). Numerous examples of ill-posed problems are given in Lattès and Lions (1967), Ivanov et al. (1978), Tikhonov and Arsenin (1979), Lavrent'ev et al. (1980), Preobrazhenskii and Pikalov (1982), and in Romanov (1984). Example 1.5 is a version of an example from Fichera (1974) which is used there in the proof of the Erling interpolation inequalities. Example 1.7 is taken from Payne (1973), Example 1.8 is a simple generalization of the construction from Taylor (1981). The problem from Example 1.7 was solved by Riesz (1938).

To Section 2. The presentation of the "well-posed part" of the section is close to Samarskii (1977). The recommended additional bibliography can be found in Bakushinskii (1971, 1972), Apartsin and Bakushinskii (1972), and Voronin and Tsetsokho (1981), where it was shown that for a certain coordination of the spacings of a difference scheme and the accuracy of initial data, the parameter h plays the role of a regularization parameter. Difference schemes for Volterra equations were investigated from this point of view in Apartsin and Bakushinskii (1977). Difference schemes for the abstract ill-posed Cauchy problem were considered by Bakushinskii (1971, 1972). Difference schemes for integral equations of the first kind of special form were studied by Voronin and Tsetsokho (1981).

To Section 3. Variational methods of solution of ill-posed problems were first investigated by Tikhonov (1963). We have presented only a formal scheme of obtaining them on the basis of the definition of l-well-posedness. The detailed investigation of this problem can be found in Morozov (1974), Ivanov et al. (1978), and in Tikhonov and Arsenin (1979). The method of choosing the regularization parameter on the basis of the residual condition (3.15) was proposed by Morozov (1966). Various methods of choosing the regularization parameter are described in the literature mentioned above and in the references therein. The variational method was first proposed by Phillips (1962) on the basis of the example of solution of the first order Fredholm equation.

Chapter 6.

Theory of stability of difference schemes

6.1. STATEMENT OF THE PROBLEM AND THE NECESSARY CONDITIONS OF FINITE STABILITY

Suppose $\mathbb{Z}$ is the set of integers, i.e., $\mathbb{Z} = \{0, \pm 1, \pm 2, \ldots, \pm n, \ldots\}$ and $u: \mathbb{Z} \to H$ is the function of $j \in \mathbb{Z}$ taking values in a real or complex Hilbert space H with norm $\|u\|$ and scalar product $\langle u, v \rangle$. In the applications to difference schemes, H is usually represented by the finite-dimensional space $\mathbb{R}^n$ or $\mathbb{C}^n$. However, the property of being finite-dimensional is not used in the proofs. Therefore, we shall consider the general case from the very beginning. Suppose τ is an arbitrary positive number. We define the difference derivatives ∂ and $\bar{\partial}$ as follows:

$$(\partial u)_j = (u_{j+1} - u_j)/\tau, \qquad (\bar{\partial} u)_j = (u_j - u_{j-1})/\tau.$$

In the following, we shall denote the dependence of the function u on the argument $j \in \mathbb{Z}$ by the index notation: $u(j) = u_j$. For these derivatives we shall also use the notation

$$\partial u = u_t, \qquad \bar{\partial} u = u_{\bar{t}}.$$

Define the operators of the shift to the right and to the left by the formulas

$$(\check{u})_j = u_{j-1}, \qquad (\hat{u})_j = u_{j+1}.$$

Now, we consider the abstract two-layer scheme with weights:

$$(Pu)_j = (u_{j+1} - u_j)/\tau - A(\sigma u_{j+1} + (1-\sigma)u_j) = f_j, \qquad (1.1)$$

$$u_0 = g, \qquad j = 0, \ldots, N-1.$$

Here A is a linear bounded operator acting in the space H, possibly, depending on j; σ is a real parameter; g and f_j are given elements of the space H; $\tau N = T$ is a constant. Using the notations introduced above, we write difference schemes (1.1) in the compact form

$$Pu = u_t - A(\sigma\hat{u} + (1-\sigma)u) = f. \qquad (1.2)$$

Below we present a typical example of the difference scheme of the form (1.2) for a partial differential equation.

Example 1.1. Suppose $u(x,t)$ is a solution of the parabolic equation with reverse time

$$\partial u/\partial t = -\partial^2 u/\partial x^2 + f(x,t), \qquad x \in \mathbb{R}, \quad t \in [0,T],$$

$$u(x,0) = 0, \qquad x \in \mathbb{R}.$$

Denote $u_j^n = u(nh, j\tau)$, where h and τ are the discretization spacings in x and t, correspondingly. We consider an elementary explicit scheme

$$(u_{j+1}^n - u_j^n)/\tau = -(u_j^{n+1} - 2u_j^n + u_j^{n-1})/h^2 + f_j^n,$$

$$u_0^n = 0. \qquad (1.3)$$

For the Hilbert space H we take the space of mesh functions $u : \mathbb{Z} \to \mathbb{R}$ with norm

$$\|u\|^2 = h \sum_{n \in \mathbb{Z}} |u^n|^2$$

and scalar product

$$\langle u, v \rangle = h \sum_{h \in \mathbb{Z}} u^n v^n.$$

We define the operator A by the formula

$$Au = -\partial_x \bar{\partial}_x u,$$

where $(\partial_x u)^n = (u^{n+1} - u^n)/h$, $(\bar{\partial}_x u)^n = (u^n - u^{n-1})/h$. Then the difference scheme (1.3) takes the form (1.2) with $\sigma = 0$, where A is a self-adjoint operator and $\|A\| = 4/h^2$.

Chapter 6. Theory of stability of difference schemes

To define finite stability we introduce the corresponding weight norms. Suppose
$$Z_0^N = \{0, 1, \ldots, N\}; \qquad \varphi : Z_0^N \to \mathbb{R}$$
is a real, monotonically decreasing weight function, i.e., $-\varphi_t > 0$. By the function φ and the number $s > 0$ we construct the function $\Psi : Z_0^{N-1} \to \mathbb{R}$

$$\Psi_t = s\hat{\Psi}\varphi_t, \qquad \Psi_0 = 1. \qquad (1.4)$$

The function Ψ is the discrete analogue of the weight function $\exp(s\varphi(t))$, whose particular case, $\varphi(t) = -t$, was already used in the previous chapter to obtain the estimate of conditional stability of the boundary value problem for the heat conduction equation with reverse time.

The analogy with exponential function is also emphasized by the following lemma.

Lemma 1.1. *The function Ψ satisfies the estimates*

$$\exp(-smj\tau) \le \Psi_j \le \exp(-s\mu j\tau/(1+s\mu\tau)) \le 1, \qquad (1.5)$$

$$\Psi_{j+1} \le \Psi_j, \qquad (1.6)$$

$$\exp(-sm(j-k)\tau) \le \Psi_j/\Psi_k \le \exp(-s\mu(j-k)\tau/(1+s\tau\mu)), \qquad 0 \le k \le j \qquad (1.7)$$

where

$$m = \max_{j=0,1,\ldots,N-1}(-\varphi_t)_j, \qquad (1.8)$$

$$\mu = \min_{j=0,1,\ldots,N-1}(-\varphi_t)_j. \qquad (1.9)$$

Proof. By (1.4) we have
$$(\Psi_{j+1} - \Psi_j)/\tau = s\varphi_{tj}\Psi_{j+1}, \qquad \Psi_0 = 1,$$
which implies inequality (1.6) (since $-s\tau\varphi_t > 0$) and also the identities

$$\Psi_j = \prod_{l=0}^{j-1}(1 - s\tau(\varphi_t)_l)^{-1}, \qquad \Psi_j/\Psi_k = \prod_{l=k}^{j-1}(1 - s\tau(\varphi_t)_l)^{-1}$$

Therefore, by the definition of the numbers m and μ, we have

$$(1 + s\tau m)^{-(j-k)} \le \Psi_j/\Psi_k \le (1 + s\tau\mu)^{-(j-k)}. \qquad (1.10)$$

It is well known that $(1+x^{-1})^x < e < (1+x^{-1})^{x+1}$ for $x > 0$. This yields

$$(1+s\tau m)^{-(j-k)} > \exp(-s\tau m(j-k)).$$

Analogously, we have

$$(1+s\tau\mu)^{-(j-k)} < \exp\{-(j-k)s\tau\mu/(1+s\tau\mu)\}.$$

Combining these estimates with (1.10), we obtain (1.7). Setting $k=0$ in (1.7), we obtain (1.5). The lemma is proved. $\square$

For the function $u: Z_0^{N-1} \to H$ we set

$$\|u\|_s^2 = \tau \sum_{j=0}^{N-1} \Psi_j^2(s)\|u_j\|^2. \tag{1.11}$$

The norm (1.11) is the discrete analogue of

$$\int_0^T \exp(2s\varphi(t))\|u(t)\|^2\,dt, \tag{1.12}$$

and (1.11) converges to (1.12) for $\tau \to 0$. If we denote by $l_2(k, N; H)$ the Hilbert space of mesh functions $u: Z_k^N \to H$, $Z_k^N = \{k, k+1, \ldots, N\}$ with norm

$$\|u\|_{l_2(k,N;H)}^2 = \tau \sum_{j=k}^N \|u_j\|^2,$$

then, by definition (1.11), we have $\|u\|_0 = \|u\|_{l_2(0,N-1;H)}$. Therefore, inequality (1.5) yields the two-sided estimate ($\tau N = T$)

$$\exp(-s(T-\tau)m)\|u\|_0 \le \|u\|_s \le \|u\|_0, \tag{1.13}$$

which shows that the norms $\|u\|_0$ and $\|u\|_s$ are equivalent.

We denote by $C_0(Z_0^N)$ the set of functions $u: Z_0^N \to H$ such that

$$u_0 = u_N = 0.$$

The linear space $C_0(Z_0^N)$ is the discrete analogue of the set $C_0(0,T)$ of finite continuous functions u in the interval $[0,T]$, such that $u(0) = u(T) = 0$.

Definition 1.1. A difference scheme P of the form (1.2) is called *stable for finite functions* (briefly, *finitely stable*) if there exist numbers s_0 and

Chapter 6. Theory of stability of difference schemes

$M > 0$ independent of τ and $\|A\|$ such that for all $s \geq s_0$, $u \in C_0(Z_0^N)$, the following estimate holds:

$$s\|u\|_s^2 \leq M\|Pu\|_s^2. \tag{1.14}$$

If estimate (1.14) were valid for any function $u : Z_0^N \to H$ with the condition $u_0 = 0$, this would mean that the difference scheme P is stable in the classical sense in the space $l_2(0, N-1; H)$. To prove this it suffices to set $s = s_0$ in (1.14) and use estimate (1.13). As a result, we obtain

$$\|u\|_0 \leq c\|Pu\|_0, \quad c = (s_0/M)^{-1/2} \exp(Tms_0).$$

The additional condition $u_N = 0$ in estimate (1.14), at first glance, does not allow using it for the investigation of conditional stability because in the difference Cauchy problem (1.1) only u_0 is given, and u_N is unknown. The following theorem shows that this is not so.

Theorem 1.1. *Suppose that the difference scheme P is finitely stable. Then there exists $\varepsilon_0 > 0$ such that for all $\varepsilon \in (0, \varepsilon_0]$, $\tau \in (0, \tau_0]$ and $u : Z_0^N \to H$ satisfying the condition $u_0 = 0$ the following estimate holds:*

$$\|u\|^2_{l_2(1,N'';H)} \leq \varepsilon^2\|(E - \sigma\tau A)\hat{u}\|^2_{l_2(N',N-1;H)} + c^2(\varepsilon)\|Pu\|^2_{l_2(0,N-1;H)}. \tag{1.15}$$

Here $1 \leq N'' \leq N' < N$, $N, N', N'' \in \mathbb{Z}$, $\tau N = T$, $\tau N' = T'$, $\tau N'' = T''$ and the function $c(\varepsilon)$ is determined implicitly from the equations

$$c(\varepsilon) = \sqrt{\frac{2M}{s}} e^{smT''}, \tag{1.16}$$

$$\varepsilon = s^{-1/2}\frac{\sqrt{2M}}{T-T'} \exp\left\{-\frac{s\mu(T'-T'')}{1+s\tau_0\mu}\right\}, \tag{1.17}$$

where the numbers m and μ are given by (1.8) and (1.9).

Proof. We define the cut-off function χ by the formula

$$\chi_j = \begin{cases} 1, & j = 0, \ldots, N', \\ (N-j)(N-N')^{-1}, & j = N', \ldots, N. \end{cases}$$

Then

$$\partial \chi_j = 0, \quad j = 0, \ldots, N'-1$$
$$\partial \chi_j = -(T-T')^{-1}, \quad j = N', \ldots, N-1.$$

The function χu belongs to $C_0(Z_0^N)$, and, consequently, estimate (1.14) holds for it:
$$s\|\chi u\|_s^2 \leq M\|P\chi u\|_s^2. \tag{1.18}$$

Now, we estimate the left-hand side of this inequality from below and the right-hand side from above. So, we shall get rid of the function χ. Since $\chi_j = 1$ and $\Psi_j \geq \Psi_{N''}$ for $j \leq N''$, we have

$$s\|\chi u\|_s^2 = s\tau \sum_{j=0}^{N-1} \Psi_j^2 \|\chi_j u_j\|^2 \geq s\tau \sum_{j=0}^{N''} \Psi_j^2 \|\chi_j u_j\|^2$$

$$= s\tau \sum_{j=0}^{N''} \Psi_j^2 \|u_j\|^2 \geq s\Psi_{N''}^2 \|u\|_{l_2(1,N'';H)}^2.$$

On the other hand, $Pu = u_t - A(\sigma\hat{u} + (1-\sigma)u) = (E - \sigma\tau A)u_t - Au$ because $\sigma\hat{u} = \sigma u + \sigma\tau u_t$. Therefore, by the formula of finite-difference differentiation of the product, $\partial(\chi u) = \chi u_t + \chi_t \hat{u}$, we have

$$P\chi u = \chi Pu + \chi_t(E - \sigma\tau A)\hat{u}.$$

Using the estimate $\|u+v\|^2 \leq 2\|u\|^2 + 2\|v\|^2$, the formula for $\partial\chi$, and the relations $|\chi_j| \leq 1$, $\Psi_j \leq 1$; $\Psi_N \leq \Psi_j$, $j \geq N'$, we obtain

$$\|P\chi u\|_s^2 \leq 2\|Pu\|_s^2 + 2\|\chi_t(E - \sigma\tau A)\hat{u}\|_s^2$$
$$\leq 2\|Pu\|_{l_2(0,N-1;H)}^2 + 2\Psi_{N'}^2(T - T')^{-2}\|(E - \sigma\tau A)\hat{u}\|_{l_2(N',N-1;H)}^2.$$

Substituting these estimates in (1.18), we find

$$s\Psi_{N''}^2 \|u\|_{l_2(1,N'';H)}^2 \leq 2M\Psi_{N'}^2(T-T')^{-2}\|(E - \sigma\tau A)\hat{u}\|_{l_2(N',N-1;H)}^2$$
$$+ 2M\|Pu\|_{l_2(0,N-1;H)}^2.$$

Due to Lemma 1.1 and estimates (1.7) and (1.5) for $\tau \leq \tau_0$, we have

$$\Psi_{N'}/\Psi_{N''} \leq \exp(-s\mu(T' - T''))(1 + s\tau_0\mu)^{-1}),$$

$$1/\Psi_{N''} \leq \exp(smT'').$$

Thus, dividing both sides of the inequality by $s\Psi_{N''}^2$ and setting

$$\varepsilon^2 = 2M(T-T')^{-2}s^{-1}\exp\left(-\frac{2s\mu(T'-T'')}{1+s\tau_0\mu}\right),$$

Chapter 6. Theory of stability of difference schemes

$$c^2(\varepsilon) = 2Ms^{-1}\exp(2smT''),$$

we obtain (1.15)–(1.17). It is easy to see that the function

$$f(s) = \sqrt{2M/s}\,(T-T')^{-1}\exp(-s\mu(T'-T'')/(1+s\tau_0\mu)), \qquad s \geq s_0$$

monotonically decreases. More precisely, $f'(s) < 0$, $\lim_{s\to\infty} f(s) = 0$. Therefore, setting $\varepsilon_0 = f(s_0)$, we see that the equation $\varepsilon = f(s)$ is uniquely solvable in the interval $(0,\varepsilon_0]$: $s = s(\varepsilon)$. This means that the function $c(\varepsilon)$ is correctly defined, which completes the proof. □

Note that estimate (1.15) becomes very simple for the explicit scheme with $\sigma = 0$:

$$\|u\|^2_{l_2(1,N'';H)} \leq \varepsilon^2 \|u\|^2_{l_2(N'+1,N;H)} + c^2(\varepsilon)\|Pu\|^2_{l_2(0,N-1;H)}. \qquad (1.19)$$

If we suppose in addition that $\|E - \sigma\tau A\| \leq d$, where the number d is independent of τ, and also in the case of the variable operator A of $j \in Z_0^{N-1}$, estimate (1.15) is reduced to (1.19) up to the multiplier d^{-1}:

$$\|u\|^2_{l_2(1,N'';H)} \leq \varepsilon^2 \|u\|^2_{l_2(N'+1,N;H)} + c^2(\varepsilon d^{-1})\|Pu\|^2_{l_2(0,N-1;H)}.$$

Theorem 1.1 shows that the condition of finite stability of the difference scheme P implies l-stability with respect to the right-hand side $f = Pu$. Here, the stabilizing functional l is defined by the formula

$$l(u) = \|(E-\sigma\tau A)\hat{u}\|_{l_2(N',N-1;H)}.$$

We now consider the difference Cauchy problem $Pv = f$, $v_0 = g$, with nonlinear initial data g. Then, by Theorem 1.1, for the function $u = v - g$ the following estimate holds:

$$\|v-g\|_{l_2(1,N'';H)} \leq \varepsilon l(v-g) + c(\varepsilon)\|P(v-g)\|_{l_2(0,N-1;H)}$$

Therefore, by the triangle inequality, we obtain

$$\|v\|_{l_2(1,N'';H)} \leq \varepsilon l(v) + \varepsilon l(g) + c(\varepsilon)\|P(v)\|_{l_2(0,N-1;H)}$$
$$+ c(\varepsilon)\|Ag\|_{l_2(0,N-1;H)} + \|g\|T''.$$

Therefore, setting

$$|g| = l(g) + T''\|g\| + \|Ag\|_{l_2(0,N-1;H)}, \qquad c_1(\varepsilon) = \max(1,\varepsilon,c(\varepsilon)),$$

we have

$$\|v\|_{l_2(1,N'';H)} \leq \varepsilon l(v) + c_1(\varepsilon)[\|Pv\|_{l_2(0,N-1;H)} + |g|]. \qquad (1.20)$$

Thus, we have established the following statement.

Theorem 1.2. *Suppose the difference scheme P is finitely stable. Then the Cauchy problem $Pv = f$, $v_0 = g$, is l-stable. More precisely, for all $\varepsilon \in (0, \varepsilon_0]$, estimate (1.20) holds.*

We shall now find the necessary conditions for finite stability, assuming that the operator A in the scheme (1.2) may depend on $j \in Z_0^{N-1}$ in the general case.

Theorem 1.3. *In order that the difference scheme (1.2) be finitely stable, it is necessary that*

$$1/\tau \notin \cup_{j=0}^{N-2} \operatorname{Sp}(\sigma A_j), \qquad (1.21)$$

where $\operatorname{Sp} A_j$ is the spectrum of the operator A_j.

Proof. Suppose that the scheme P is finitely stable, i.e., for any $s \geq s_0$ and $u \in C_0(Z_0^N)$ we have

$$s\|u\|_s^2 \leq M\|u_t - A(\sigma\hat{u} + (1-\sigma)u)\|_s^2. \qquad (1.22)$$

We set $\Psi u = v$. By the formula of finite-difference differentiation of a product, we have

$$u_t = (\Psi^{-1}v)_t = (\Psi^{-1})_t \hat{v} + \Psi^{-1} v_t.$$

From (1.4) it follows that $(1/\Psi)_t = (-s(\varphi_t)_j)/\Psi_j$ and therefore

$$u_t = (v_t - s\varphi_t \hat{v})/\Psi. \qquad (1.23)$$

Since $\Psi/\hat{\Psi} = 1 - s\tau\varphi_t$, we have

$$\sigma\hat{u} + (1-\sigma)u = \frac{1}{\Psi}\{\sigma(1 - s\tau\varphi_t)\hat{v} + (1-\sigma)v\} \equiv \frac{v_\sigma}{\Psi} \qquad (1.24)$$

(the expression in braces is assumed as the definition of the function v_σ). As a result, after the change of variables, estimate (1.22) on account of the definition (1.11) takes the form

$$s\tau \sum_{j=0}^{N-1} \|v_j\|^2 \leq \tau M \sum_{j=0}^{N-1} \|(v_t - s\varphi_t \hat{v} - Av_\sigma)_j\|^2. \qquad (1.25)$$

Set $v_j = \delta_{jk} w$, where k is fixed, $k \in Z_1^{N-1}$, δ_{jk} is the Kronecker symbol, and w is the eigenvector of the operator A_{k-1}: $A_{k-1} w = \lambda w$, $\|w\| = 1$.

Since

$$(v_t)_j = \begin{cases} -w/\tau, & j = k, \\ w/\tau, & j = k-1, \\ 0, & j \neq k, k-1; \end{cases}$$

$$(v_\sigma)_j = \begin{cases} (1-\sigma)w, & j = k \\ \sigma(1 - s\tau\varphi_t)w, & j = k-1 \\ 0, & j \neq k, k-1, \end{cases}$$

after the substitution of this function v in estimate (1.25) and dividing by τ, we obtain

$$s \leq M\{\|-w/\tau - (1-\sigma)A_k w\|^2 + \|w/\tau - s(\varphi_t)_{k-1} w - \lambda\sigma(1 - s\tau(\varphi_t)_{k-1})w\|^2\}.$$

Hence,

$$s \leq M\Big\{\|w/\tau + (1-\sigma)A_k w\|^2 + |\tau^{-1} - \lambda\sigma + s(\varphi_t)_{k-1}(\lambda\sigma\tau - 1)|^2\Big\}.$$

For $\lambda\sigma\tau = 1$, the right-hand side of the inequality is independent of s and the left-hand side tends to infinity as $s \to \infty$. This contradiction proves the theorem in the case where the spectrum of the operator is purely discrete. If the space H is infinite-dimensional, then $\lambda = 1/\tau\sigma$ may be a point of continuous spectrum of the operator A_{k-1}. In this case, for any natural $n \geq 1$ there exists an element $w_n \in H$ such that

$$\|(A_{k-1} - \lambda)w_n\| \leq 1/n, \qquad \|w_n\| = 1.$$

Setting $v = \delta_{jk} w_n$ in (1.25), we have

$$s \leq M\{\|w_n/\tau + (1-\sigma)A_k w_n\|^2 + 2|\tau^{-1} - \lambda\sigma + s(\varphi_t)_{k-1}(\lambda\sigma\tau - 1)|^2 \\ + 2\|\sigma(1 - s\tau\varphi_{t\ k-1})(A_{k-1} - \lambda)w_n\|^2\} \\ \leq M\{(\tau^{-1} + |1-\sigma|\,\|A_k\|)^2 + 2|\tau^{-1} - \lambda\sigma + s(\varphi_t)_{k-1}(\lambda\sigma\tau - 1)|^2 \\ + 2\sigma|1 - s\tau(\varphi_t)_{k-1}|^2 n^{-2}\}.$$

For $\lambda\sigma\tau = 1$, passing to the limit, first, as $n \to \infty$, and then as $s \to \infty$, we arrive at a contradiction again. So, we have shown that if $\tau^{-1} = \sigma\lambda$, $\lambda \in \operatorname{Sp} A_{k-1}$, $k = 1, \ldots, N-1$, then estimate (1.22) is invalid. The theorem is proved. □

Note that condition (1.21) is nontrivial even for the simplest case $A = A^* > 0$, the operator A is independent of j, the space H is finite-dimensional, and $\sigma \leq 1$. Indeed, for $f = 0$ the equation $Pu = f$ becomes as follows:

$$(E - \sigma\tau A)u_{j+1} = (E + \tau(1-\sigma)A)u_j. \tag{1.26}$$

If $u_0 = 0$, then, evidently, for the unique determination of $u_1, u_2, \ldots, u_N$ from the recurrent relation (1.26) it is necessary that $\det(E - \sigma\tau A) \neq 0$, i.e., $\tau^{-1} \notin \mathrm{Sp}(\sigma A)$. However, u_N is also equal to zero and therefore, since $\det(E + \tau(1-\sigma)A) \neq 0$, we may perform calculations in reverse direction and uniquely determine $u_{N-1}, u_{N-2}, \ldots, u_1$. Thus, we have proved that $\mathrm{Ker}\, P = \{0\}$, where the operator

$$P : l_2(0, N; H) \cap C_0(Z_0^N) \to l_2(0, N-1; H)$$

is defined by formula (1.2). Since the space $l_2(0, N-1; H)$ is finite-dimensional and its norm is equivalent to the weight norm $\|u\|_s$ (see (1.13)), the following estimate holds:

$$s\|u\|_s \leq M\|Pu\|_s.$$

However, in the general case, the constant M may depend on s, τ, and $\|A\|$.

Exercise 1.1. Suppose the difference scheme P of the form (1.2) is finitely stable and Q is a linear operator in $l_2(0, N-1; H)$ such that

$$\|Qu\|_{l_2(0,N-1;H)} \leq c\|u\|_{l_2(0,N-1;H)} \tag{1.27}$$

for all $u \in C_0(Z_0^N)$. Show that the difference scheme $P + Q$ is also finitely stable, i.e., for all $u \in C_0(Z_0^N)$

$$s\|u\|_s^2 \leq M\|(P+Q)u\|_s^2, \qquad s \geq s_0,$$

where M is independent of s, τ, and $\|A\|$.

Exercise 1.2. Prove the above statement with condition (1.27) replaced by the estimate

$$\|Qu\|_s^2 \leq s^\alpha c\|u\|_s^2, \qquad u \in C_0(Z_0^N), \quad \alpha < 1.$$

Exercise 1.3. Suppose the operator A and the space H are the same as in Example 1.1. Find $\mathrm{Sp}(\sigma A)$ and specify the set of τ for which the scheme with weights, (1.2), can be finitely stable.

6.2. BASIC ESTIMATES

To obtain the sufficient conditions of finite stability, we need to estimate the quantity

$$I \equiv \|Pu\|_s^2$$

from below. Every operator of $\mathcal{L}(H)$ can be represented as a sum of self-adjoint and skew-self-adjoint terms:

$$A = \frac{A + A^*}{2} + i\frac{(A - A^*)}{2i}, \qquad i^2 = -1.$$

So, it will be convenient for us to write $A + iB$ instead of the operator A in formula (1.2), assuming that $A = A^*$ and $B = B^*$. Thus, in this section we have

$$Pu = u_t - (A + iB)(\sigma\hat{u} + (1 - \sigma)u).$$

In the general case, A and B do not commute and depend on $j \in Z_0^N$. Further, we shall need the following conditions on the difference derivative $\partial A = A_t$ and the commutator $[A, B] = AB - BA$:

$$-A_t \leq cA, \tag{2.1}$$

$$-i[A, B] \leq c_1 A. \tag{2.2}$$

Recall that $A \geq D$ means that

$$\langle Au, u \rangle \geq \langle Du, u \rangle, \qquad \forall u \in H.$$

Making the change of variables $v = \Psi u$ and using formulas (1.23) and (1.24), we obtain

$$I = \tau \sum_{j=0}^{N-1} \|v_t - iBv_\sigma - Av_\sigma - s\varphi_t\hat{v}\|^2$$

$$= \tau \sum_{j=0}^{N-1} \{\|v_t - iBv_\sigma\|^2 + \|Av_\sigma + s\varphi_t\hat{v}\|^2 - 2\operatorname{Re}\langle v_t, s\varphi_t\hat{v}\rangle$$

$$+ 2\operatorname{Re}\langle iBv_\sigma, s\varphi_t\hat{v}\rangle - 2\operatorname{Re}\langle v_t, Av_\sigma\rangle + 2\operatorname{Re}\langle iBv_\sigma, Av_\sigma\rangle\}$$

$$\equiv \sum_{k=1}^{6} I_k. \tag{2.3}$$

Here we denote by I_k the sum $\tau \sum_{j=0}^{N-1}$ corresponding to the kth term in braces in the expression (2.3). We introduce the following notation for numerical functions of discrete argument.

$$[x,y] \equiv \tau \sum_{j=0}^{N-1} x_j y_j, \qquad (2.4)$$

$$(x,y) \equiv \tau \sum_{j=1}^{N-1} x_j y_j, \qquad (2.5)$$

and estimate every term I_k starting from I_3. For functions of discrete argument $v, w : \mathbb{Z} \to H$, the formula of finite-difference differentiation of a scalar product is as follows:

$$\partial\langle v, w\rangle = \langle v_t, \hat{w}\rangle + \langle v, w_t\rangle. \qquad (2.6)$$

In particular, for $w = v$,

$$\partial\|v\|^2 = \langle v_t, \hat{v}\rangle + \langle v - \hat{v}, v_t\rangle + \langle \hat{v}, v_t\rangle = -\tau\|v_t\|^2 + 2\operatorname{Re}\langle v_t, \hat{v}\rangle,$$

i.e.,

$$2\operatorname{Re}\langle v_t, \hat{v}\rangle = \partial\|v\|^2 + \tau\|\partial v\|^2. \qquad (2.7)$$

Hence,

$$I_3 \equiv -s[\varphi_t, 2\operatorname{Re}\langle v_t, \hat{v}\rangle] = -s[\varphi_t, \partial\|v\|^2] - \tau s[\varphi_t, \|\partial v\|^2].$$

By the formula of summing by parts, we obtain

$$[x, \partial y] = -(\bar{\partial}x, y) + (x_{N-1}y_N - x_0 y_0) \qquad (2.8)$$

Therefore, transferring the operator ∂ in the first term I_3 to φ_t and taking into account the condition $v_0 = v_N = 0$, we obtain

$$I_3 = s(\varphi_{t\bar{t}}, \|v\|^2) - \tau s[\varphi_t, \|\partial v\|^2].$$

Since $v_\sigma = \sigma(1 - s\tau\varphi_t)\hat{v} + (1-\sigma)v$, we have

$$2\operatorname{Re}\langle iBv_\sigma, \hat{v}\rangle = \sigma(1 - s\tau\varphi_t)\, 2\operatorname{Re}\langle iB\hat{v}, \hat{v}\rangle$$
$$+ 2(1-\sigma)\operatorname{Re}\langle iBv, \hat{v} - v\rangle + 2(1-\sigma)\operatorname{Re}\langle iBv, v\rangle.$$

The operator B is self-adjoint and therefore $\operatorname{Re}\langle iBv, v\rangle = 0$ for all $v \in H$. Consequently, $2\operatorname{Re}\langle iBv_\sigma, \hat{v}\rangle = 2(1-\sigma)\tau\operatorname{Re}\langle iBv, v_t\rangle$ and

$$|2\operatorname{Re}\langle iBv_\sigma, \hat{v}\rangle| \le \tau|1 - \sigma|\{\alpha\|B\|^2\|v\|^2 + \alpha^{-1}\|v_t\|^2\}.$$

Chapter 6. Theory of stability of difference schemes

Here we have used the inequality $2ab \leq \alpha a^2 + \alpha^{-1} b^2$, $\alpha > 0$. Since $I_4 = s[\varphi_t, 2\operatorname{Re}\langle iBv_\sigma, \hat{v}\rangle)$, we obtain

$$|I_4| \leq |1 - \sigma|\tau s\left[\,|\varphi_t|, \alpha \|B\|^2 \|v\|^2\right) + |1 - \sigma|\tau s\left[\,|\varphi_t|, \alpha^{-1}\|v_t\|^2\right).$$

We now estimate I_5. By the definition of v_σ, we have

$$\begin{aligned}
2\operatorname{Re}\langle v_t, Av_\sigma\rangle &= \sigma(1 - s\tau\varphi_t)2\operatorname{Re}\langle v_t, A\hat{v}\rangle + (1-\sigma)2\operatorname{Re}\langle v_t, Av\rangle \\
&= \sigma(1 - s\tau\varphi_t)2\operatorname{Re}\langle v_t, A(\hat{v} - v + v)\rangle + (1-\sigma)2\operatorname{Re}\langle v_t, Av\rangle \\
&= \tau\sigma(1 - s\tau\varphi_t)2\langle v_t, Av_t\rangle + (1 - s\tau\varphi_t\sigma)2\operatorname{Re}\langle v_t, Av\rangle. \quad (2.9)
\end{aligned}$$

To transform the last term, we note that

$$\partial\langle v, Av\rangle = \langle v_t, Av\rangle + \langle \hat{v}, (Av)_{\bar{t}}\rangle, \qquad (Av)_t = A_t\hat{v} + Av_t.$$

Hence,

$$\begin{aligned}
\partial\langle v, Av\rangle &= \langle v_t, Av\rangle + \langle \hat{v}, A_t\hat{v}\rangle + \langle \hat{v} - v + v, Av_t\rangle \\
&= \langle v_t, Av\rangle + \langle \hat{v}, A_t\hat{v}\rangle + \tau\langle v_t, Av_t\rangle + \langle v, Av_t\rangle.
\end{aligned}$$

Since $2\operatorname{Re}\langle v_t, Av\rangle = \langle v_t, Av\rangle + \langle v, Av_t\rangle$, $A = A^*$, we get

$$2\operatorname{Re}\langle v_t, Av\rangle = \partial\langle v, Av\rangle - \langle \hat{v}, A_t\hat{v}\rangle - \tau\langle v_t, Av_t\rangle.$$

Substituting this formula into (2.9), we get

$$\begin{aligned}
2\operatorname{Re}\langle v_t, Av_\sigma\rangle = (1 - s\sigma\tau\varphi_t)\partial\langle v, Av\rangle &- (1 - s\sigma\tau\varphi_t)\langle \hat{v}, A_t\hat{v}\rangle \\
&+ \tau[2\sigma - 1 - \sigma s\tau\varphi_t]\langle v_t, Av_t\rangle. \quad (2.10)
\end{aligned}$$

By the condition (2.1) we have $-A_t \leq cA$. Consequently,

$$\begin{aligned}
-(1 - s\sigma\tau\varphi_t)&\langle \hat{v}, A_t\hat{v}\rangle \\
&\leq c(1 - s\sigma\tau\varphi_t)\langle \hat{v}, A\hat{v}\rangle \\
&= c\operatorname{Re}\langle \hat{v}, A[(1 - s\sigma\tau\varphi_t)\hat{v} + \sigma\hat{v} - \sigma\hat{v} + (1-\sigma)v - (1-\sigma)v]\rangle \\
&= c\operatorname{Re}\langle \hat{v}, Av_\sigma\rangle + c\operatorname{Re}\langle \hat{v}, Av_t\rangle\tau(1-\sigma). \quad (2.11)
\end{aligned}$$

Here we have used the condition $1 - s\sigma\tau\varphi_t \geq 0$. For this condition to hold, we shall assume further that $\sigma \geq 0$ and $\varphi_t \leq 0$.

Note that

$$\begin{aligned}
\operatorname{Re}\langle \hat{v}, Av_t\rangle &= \operatorname{Re}\langle \sigma(1 - s\tau\varphi_t)\hat{v} + \hat{v} - \sigma\hat{v} + \sigma s\tau\varphi_t\hat{v} + (1-\sigma)v \\
&\qquad - (1-\sigma)v, Av_t\rangle \\
&= \operatorname{Re}\langle v_\sigma, Av_t\rangle + \tau(1-\sigma)\langle v_t, Av_t\rangle + \sigma s\tau\varphi_t\operatorname{Re}\langle \hat{v}, Av_t\rangle.
\end{aligned}$$

Hence,
$$\mathrm{Re}\langle \hat{v}, Av_t\rangle = (1 - \sigma s\tau\varphi_t)^{-1}\{\mathrm{Re}\langle v_\sigma, Av_t\rangle + \tau(1-\sigma)\langle v_t, Av_t\rangle\}.$$

Substituting this formula into estimate (2.11) and taking into account that
$$\mathrm{Re}\langle \hat{v}, Av_\sigma\rangle = \mathrm{Re}\{\langle v, Av_\sigma\rangle + \tau\langle v_t, Av_\sigma\rangle\},$$

we have
$$-(1 - s\sigma\tau\varphi_t)\langle \hat{v}, A_t\hat{v}\rangle$$
$$\le c\,\mathrm{Re}\langle v, Av_\sigma\rangle + [\tau c + c\tau(1-\sigma)(1-\sigma s\tau\varphi_t)^{-1}]\,\mathrm{Re}\langle v_t, Av_\sigma\rangle$$
$$+ c\tau^2(1-\sigma)^2(1-\sigma s\tau\varphi_t)^{-1}\langle v_t, Av_t\rangle. \tag{2.12}$$

Estimate the term $\mathrm{Re}\langle v, Av_\sigma\rangle$:
$$\mathrm{Re}\langle v, Av_\sigma\rangle = \mathrm{Re}\langle v, Av_\sigma + s\varphi_t\hat{v} - s\varphi_t\hat{v}\rangle = \mathrm{Re}\{\langle v, Av_\sigma + s\varphi_t\hat{v}\rangle - s\varphi_t\langle v, \hat{v}\rangle$$
$$\le \alpha\|Av_\sigma + s\varphi_t\hat{v}\|^2/2 + \|v\|^2/2\alpha - s\varphi_t\|v\|^2/2 - s\varphi_t\|\hat{v}\|^2/2\}. \tag{2.13}$$

Here we have used the inequality $2ab \le \alpha a^2 + \alpha^{-1}b^2$, $\alpha > 0$, and the condition $-\varphi_t \ge 0$. Combining estimates (2.10)–(2.13), we have
$$-2\,\mathrm{Re}\langle v_t, Av_\sigma\rangle = -(1 - s\sigma\tau\varphi_t)\partial\langle v, Av\rangle + (1 - s\sigma\tau\varphi_t)\langle \hat{v}, A_t\hat{v}\rangle$$
$$+ \tau(1 - 2\sigma + \sigma s\tau\varphi_t)\langle v_t, Av_t\rangle$$
$$\ge -(1 - s\sigma\tau\varphi_t)\partial\langle v, Av\rangle + \tau(1 - 2\sigma + \sigma s\tau\varphi_t)\langle v_t, Av_t\rangle$$
$$- [\tau c + c\tau(1-\sigma)(1-\sigma s\tau\varphi_t)^{-1}]\,\mathrm{Re}\langle v_t, Av_\sigma\rangle$$
$$- c\tau^2(1-\sigma)^2(1-\sigma s\tau\varphi_t)^{-1}\langle v_t, Av_t\rangle - c\alpha\|Av_\sigma + s\varphi_t\hat{v}\|^2/2$$
$$- c\|v\|^2/2\alpha + cs\varphi_t\|v\|^2/2 + sc\varphi_t\|\hat{v}\|^2/2.$$

Collecting the like terms and multiplying this inequality by the function
$$\lambda = \{1 - [\tau c + \tau c(1-\sigma)(1-\sigma s\tau\varphi_t)^{-1}]/2\}^{-1},$$

which is positive for sufficiently small $\tau < \tau_0$, we obtain
$$-2\,\mathrm{Re}\langle v_t, Av_\sigma\rangle \ge -\lambda(1 - s\sigma\tau\varphi_t)\partial\langle v, Av\rangle + \lambda\langle v_t, Av_t\rangle\{\tau(1 - 2\sigma + \sigma s\tau\varphi_t)$$
$$- c\tau^2(1-\sigma)^2(1-\sigma s\tau\varphi_t)^{-1}\} - \bar{\lambda}c\alpha\|Av_\sigma + s\varphi_t\hat{v}\|^2/2$$
$$- \bar{\lambda}c(1/\alpha - s\varphi_t)\|v\|^2/2 + sc\bar{\lambda}\varphi_t\|\hat{v}\|^2/2. \tag{2.14}$$

Chapter 6. Theory of stability of difference schemes

Here $\bar{\lambda} = \max_j \lambda$. Hence, taking into account the identities

$$[\varphi_t, \|\hat{v}\|) = (\varphi_{\bar{t}}, \|v\|), \qquad v \in C_0(Z_0^N),$$

$$-[\lambda(1 - s\sigma\tau\varphi_t), \partial\langle v, Av\rangle) = (\bar{\partial}(\lambda(1 - \sigma s\tau\varphi_t)), \langle v, Av\rangle)$$
$$= (-\lambda s\sigma\tau\varphi_{t\bar{t}}, \langle v, Av\rangle) + (\lambda_{\bar{t}}(1 - s\sigma\tau\varphi_t), \langle v, Av\rangle), \tag{2.15}$$

from inequality (2.14) and the definition of I_5 and I_2, we get

$$I_5 = [1, -2\operatorname{Re}\langle v_t, Av_\sigma\rangle)$$
$$\geq -\lambda s\sigma\tau(\varphi_{t\bar{t}}, \langle v, Av\rangle) + (\lambda_{\bar{t}}(1 - s\sigma\tau\varphi_t), \langle v, Av\rangle)$$
$$+ [\lambda\{\tau(1 - 2\sigma + \sigma s\tau\varphi_t) - c\tau^2(1 - \sigma)^2(1 - \sigma s\tau\varphi_t)^{-1}\}, \langle v_t, Av_t\rangle)$$
$$- \bar{\lambda}c\alpha/2\, I_2 - \bar{\lambda}c(1/2\alpha - s\varphi_t/2, \|v\|^2) + \bar{\lambda}cs/2\,(\varphi_t, \|v\|^2).$$

Note that the most tedious part of the estimate I_5 is connected with the dependence of the operator A from j. If $A_j \equiv 0$, i.e., the operator A is constant, then formulas (2.10) and (2.15) with $\lambda \equiv 1$ yield

$$I_5 = -s\sigma\tau(\varphi_{t\bar{t}}, \langle v, Av\rangle) + \tau[1 - 2\sigma + \sigma s\tau\varphi_t, \langle v_{\bar{t}}, Av_t\rangle).$$

Before estimating I_6, we observe that

$$2\operatorname{Re}\langle iBv_\sigma, Av_\sigma\rangle = \langle iBv_\sigma, Av_\sigma\rangle + \langle Av_\sigma, iBv_\sigma\rangle = \langle i[A, B]v_\sigma, v_\sigma\rangle,$$

which implies $I_6 = [1, \langle i[A, B]v_\sigma, v_\sigma\rangle)$. Therefore, $I_6 \geq 0$ for $i[A, B] \geq 0$. In particular, $I_6 = 0$ if the operators A and B commute.

We now consider the general case assuming that (see (2.2)) $i[A, B] \geq -c_1 A$ and $\sigma = 0$. Then $v_\sigma = v$ and

$$\langle i[A, B]v, v\rangle \geq -c_1\langle Av, v\rangle = -c_1\langle Av + s\varphi_t\hat{v} - s\varphi_t\hat{v}, v\rangle$$
$$= -c_1\langle Av + s\varphi_t\hat{v}, v\rangle + c_1 s\varphi_t\langle \hat{v}, v\rangle$$
$$\geq -c_1\alpha\|Av + s\varphi_t\hat{v}\|^2/2 - c_1\|v\|^2/2\alpha + c_1 s\varphi_t\|v\|^2/2$$
$$+ c_1 s\varphi_t\|\hat{v}\|^2/2.$$

Therefore,

$$I_6 \geq -c_1\alpha I_2/2 - c_1(\{1/2\alpha - s\varphi_t/2 - s\varphi_{\bar{t}}/2\}, \|v\|^2).$$

Now, we sum up our results. The above estimates yield the following lemma.

Lemma 2.1. *Suppose that* $Pu = u_t - (A + iB)(\sigma \hat{u} + (1-\sigma)u)$, *where* $A = A^*$, $B = B^*$, $-A_t \leq cA$, *and* $-i[A,B] \leq c_1 A$. *Then the following representation is valid for all* $u \in C_0(Z_0^N)$:

$$\|Pu\|_s^2 = \sum_{k=1}^{6} I_k,$$

$I_1 = [1, \|v_t - iBv_\sigma\|^2] \geq 0, \quad I_2 = [1, \|Av_\sigma + s\varphi_t \hat{v}\|^2] \geq 0,$
$I_3 = (1, s\varphi_{t\bar{t}}\|v\|^2) + [1, -\tau s \varphi_t \|v_t\|^2],$
$I_4 \geq |1 - \sigma|\tau s\{[1, \varphi_t \alpha \|B\|^2 \|v\|^2] + [1, \varphi_t \alpha^{-1} \|v_t\|^2]\},$
$I_5 \geq (1, \langle \Lambda v, v \rangle) + [1, \langle D v_t, v_t \rangle] - \bar{\lambda}c\beta I_2/2,$

$\Lambda = \{-\lambda s \sigma \tau \varphi_{t\bar{t}} + \lambda_{\bar{t}}(1 - \sigma\tau s\varphi_t)\}A + \{\bar{\lambda}cs(\varphi_t + \varphi_{\bar{t}})/2 - \bar{\lambda}c/2\beta\}E, \quad (2.16)$

$\lambda = \{1 - [\tau c + \tau c(1-\sigma)/(1 - \sigma\tau s\varphi_t)]/2\}^{-1} \quad (2.17)$

$\bar{\lambda} = \max_j \lambda,$

$D = \tau\lambda\{1 - 2\sigma + \sigma s \tau \varphi_t - c\tau(1-\sigma)^2/(1 - \sigma s \tau \varphi_t)\}A, \quad (2.18)$

$I_6 \geq 0 \quad \text{for} \quad c_1 = 0, \quad \sigma \in [0,1],$

$I_6 \geq -\frac{c_1 \gamma}{2} I_2 - [1, \langle Lv, v\rangle], \quad (2.19)$

$L = c_1 \left(\frac{\gamma}{2} - s\frac{\varphi_t + \varphi_{\bar{t}}}{2} \right) E,$

for $\sigma = 0$, $c_1 \geq 0$. *Here* $v = \Psi u$, $\Psi_t = s\hat{\Psi}\varphi_t$,

$\varphi_t \leq 0, \quad v_\sigma = \sigma(1 - s\tau\varphi_t)\hat{v} + (1 - \sigma)v, \quad (2.20)$

$\bar{\lambda}c\beta/2 + c_1\gamma/2 = 1/2.$

This lemma implies the following theorem.

Theorem 2.1. *Under the hypotheses of Lemma 2.1, suppose that for all* $s \geq s_0$ *and some* $\delta > 0$

$$M_1 \equiv \left[s\varphi_{t\bar{t}} + |1-\sigma|\tau s \varphi_t \alpha \|B\|^2\right] E + \Lambda - L \geq s\delta E, \quad (2.21)$$

$$M_0 \equiv -s\tau\varphi_t(1 - |1-\sigma|\alpha^{-1})E + D \geq 0. \quad (2.22)$$

Then, for all $u \in C_0(Z_0^N)$, $s \geq s_0$, the following estimate holds:

$$s\|u\|_s^2 + I_2/2\delta \leq \delta^{-1}\|Pu\|_s^2 \qquad (2.23)$$

If $B = 0$, then this estimate holds under the condition that

$$N_1 \equiv s\varphi_{t\bar{t}}E + \Lambda \geq s\delta E, \qquad (2.24)$$

$$N_0 \equiv (1 - \tau s\varphi_t)E + D \geq 0. \qquad (2.25)$$

Proof. Omitting I_1 and collecting the terms with v and v_t separately, by Lemma 2.1 we obtain

$$\|Pu\|_s^2 \geq I_2(1 - \bar{\lambda}c\beta/2 - c_1\gamma/2) + (1, \langle M_1 v, v\rangle) + [1, \langle M_0 v_t, v_t\rangle].$$

Hence, taking into account conditions (2.20)–(2.22), after dividing by δ we obtain estimate (2.23). The case $B = 0$ is considered analogously, but the term $I_1 = [1, \|v_t\|^2]$ is not omitted and $I_4 \equiv 0$. The theorem is proved. $\square$

In the next section we transform conditions (2.21)–(2.25) to the simple criterions which can be easily verified.

6.3. SUFFICIENT STABILITY CONDITIONS

We begin with an elementary case where A is a constant operator, $A = A^*$, and $B = 0$. Then, in Lemma 2.1, $I_4 = I_6 = 0$; $\lambda = 1$, $c = c_1 = 0$. Further, we shall assume that

$$\varphi_{t\bar{t}} \geq 1, \qquad -\varphi_t \geq 1, \qquad \sigma \in \mathbb{R}. \qquad (3.1)$$

Set

$$Pu = u_t - A(\sigma\hat{u} + (1 - \sigma)u).$$

Theorem 3.1. *Suppose*

$$E - \sigma\tau A \geq \delta E, \qquad \delta > 0,$$

$$E + \tau(1 - 2\sigma)A \geq 0.$$

Then the estimate

$$s\|u\|_s^2 \leq \delta^{-1}\|Pu\|_s^2 \qquad (3.2)$$

holds for all $s > 0$, $u \in C_0(Z_0^N)$.

Proof. Lemma 2.1 and estimates (2.24) and (2.25) yield

$$N_1 = s\varphi_{t\bar{t}}(E - \sigma\tau A) \geq s\varphi_{t\bar{t}}\delta E \geq s\delta E,$$

$$N_0 = E + \tau(1 - 2\sigma)A - s\varphi_t\tau(E - \sigma\tau A) \geq 0. \qquad (3.3)$$

Therefore, the reference to Theorem 2.1 finishes the proof. □

Theorem 3.2. *Suppose $A = A^* \geq 0$, $\sigma\tau\|A\| \leq q < 1$. Then the difference scheme P is finitely stable. If $\sigma\tau\|A\| = 1$, then the scheme is not finitely stable.*

Proof. Evidently, $E - \sigma\tau A \geq \delta E$, where $\delta = 1 - q > 0$. The operator N_0 from (3.3) can be written as follows

$$N = (1 - s\varphi_t\tau)(E - \sigma\tau A) + \tau(1 - \sigma)A \geq 0;$$

therefore, as in Theorem 3.1, we obtain estimate (3.2). If $\sigma\tau\|A\| = 1$, then $1/\tau \in \mathrm{Sp}\,\sigma A$ and by Theorem 1.3 the scheme with weights is finitely unstable. The theorem is proved. □

Theorem 3.3. *Suppose $A = A^* \leq 0$ and $E + \tau(1 - 2\sigma)A \geq 0$. Then the estimate*

$$s\|u\|_s^2 \leq \|Pu\|_s^2$$

holds for all $u : Z_0^N \to H$, $u_0 = 0$. In particular, this estimate holds for

$$\sigma \geq (1 - 1/\tau\|A\|)/2. \qquad (3.4)$$

Proof. It is easy to see that if $A \leq 0$, then the condition $u_N = 0$ in Lemma 2.1 can be omitted because the additional term in the estimates of I_3 and I_5 arising when integrating by parts are nonnegative. Therefore, Theorem 3.3 follows from Theorem 3.1, where we must set $\delta = 1$ and $A \leq 0$. Since $E \geq -A/\|A\|$,

$$E + \tau(1 - 2\sigma)A \geq (1/\|A\| + \tau(1 - 2\sigma))A \geq 0$$

under condition (3.4). The theorem is proved. □

Remark 3.1. Condition (3.4) is stronger than the known condition (Samarskii, 1977, p. 361)

$$\sigma \geq 1/2 - 1/\tau\|A\|,$$

Chapter 6. Theory of stability of difference schemes

which is necessary and sufficient for the stability of the scheme with weights in the energy norm with constant 1. This fact is connected with the use of different definition of stability.

Theorems 3.2, 3.3 show that in the ill-posed case $A \geq 0$ without the condition for the mesh spacing only the explicit scheme is stable. In the well-posed case $A \leq 0$, only the implicit scheme is stable. Since in the general case of ill-posed problem the operator $A = A_+ + A_-$ may contain the positive part $A_+ \geq 0$ as well as the negative one, $A_- \leq 0$, to construct the unconditionally stable scheme we can consider the scheme with two weights $\sigma, q \in [0,1]$

$$Pu \equiv u_t - A_+(\sigma \hat{u} + (1-\sigma)u) - A_-(q\hat{u} + (1-q)u). \qquad (3.5)$$

In this case, similarly to the proof of Lemma 2.1, we have $(v = \Psi u)$

$$I \equiv \|Pu\|_s^2 = \tau \sum_{j=0}^{N-1} \|v_t - A_+ v_\sigma - A_- v_q - s\varphi_t \hat{v}\|^2$$

$$= \tau \sum_{j=0}^{N-1} \{\|v_t\|^2 + \|A_+ v_\sigma + A_- v_q - s\varphi_t \hat{v}\|^2 - 2\operatorname{Re}\langle v_t, s\varphi_t \hat{v}\rangle$$

$$- 2\operatorname{Re}\langle v_t, A_+ v_\sigma\rangle - 2\operatorname{Re}\langle v_t, A_- v_q\rangle\} \equiv \sum_{1}^{5} I_k,$$

where we have

$$I_1 = [1, \|v_t\|^2), \qquad I_2 \geq 0,$$
$$I_3 = (s\varphi_{t\bar{t}}, \|v\|^2) - [\tau s \varphi_t, \|v_t\|^2),$$
$$I_4 = (1, \langle -s\sigma\tau\varphi_{t\bar{t}} A_+ v, v\rangle) + [1, \langle \{\tau(1-2\sigma + \sigma s\tau\varphi_t)\}A_+ v_t, v_t\rangle),$$
$$I_5 = (1, \langle -qs\tau\varphi_{t\bar{t}} A_- v, v\rangle) + [1, \langle \tau(1-2q + qs\tau\varphi_t)A_- v_t, v_t\rangle).$$

Thus,

$$I \geq (1, \langle N_1 v, v\rangle) + [1, \langle N_0 v_t, v_t\rangle),$$

$$N_1 = s\varphi_{t\bar{t}}(E - \sigma \tau A_+ - q\tau A_-),$$

$$N_0 = E + \tau(1-2\sigma)A_+ + \tau(1-2q)A_- - s\tau\varphi_t(E - \sigma\tau A_+ - q\tau A_-)$$

and therefore the following theorem holds, which is analogous to Theorem 3.1.

Theorem 3.4. Let
$$E - \sigma\tau A_+ - q\tau A_- \geq \delta E, \qquad \delta > 0;$$
$$E + \tau(1 - 2\sigma)A_+ + \tau(1 - 2q)A_- \geq 0.$$

Then, for the difference scheme (3.5) estimate (3.2) holds. In particular, the scheme (3.5) is unconditionally finitely stable for $\sigma = 0$ and $q = 1$.

Following Samarskii (1977), the two-layer scheme can be written in the canonical form
$$Pu = Bu_t + Au = f. \tag{3.6}$$
Assuming that $A = A^*$, $B = B^* > 0$, $[A, B] = 0$, we obtain the following theorem.

Theorem 3.5. Suppose $\|B^{-1}\| \leq c$ and
$$B \geq \tau A. \tag{3.7}$$

Then the scheme (3.6) is finitely stable. More precisely, for all $s > 0$, $u \in C_0(Z_0^N)$, we have
$$s\|u\|_s^2 \leq c^2 \|Pu\|_s^2. \tag{3.8}$$

Proof. Rewrite (3.6) in the form
$$u_t - (-B^{-1}A)u = B^{-1}Pu.$$
Applying Theorem 3.1 with $\sigma = 0$ and $\delta = 1$, under the condition $E + \tau(-B^{-1}A) \geq 0$ we have the following estimate:
$$s\|u\|_s^2 \leq \|B^{-1}Pu\|_s^2. \tag{3.9}$$
Since $B > 0$, the condition $E + \tau(-B^{-1}A) \geq 0$ is equivalent to condition (3.7). Observing that B^{-1} is bounded, we obtain estimate (3.8). □

Corollary 3.1. If $A \geq 0$ under the assumptions of Theorem 3.5, then estimate (3.9) holds for all $u : Z_0^N \to H$ such that $u_0 = 0$.

The proof of this corollary is similar to the proof of Theorem 3.3.

The meaning of condition (3.7) is clear from an elementary example. We consider the difference scheme
$$b(u_{j+1} - u_j)/\tau + au_j = 0, \quad j = 0, 1, \ldots, \qquad u_0 = 1,$$

Chapter 6. Theory of stability of difference schemes

where $b > 0$. The corresponding differential equation

$$\frac{b\,dv}{dt} + av = 0, \quad t > 0, \quad v(0) = 1$$

has the solution $v(t) = \exp(-at/b)$ and, consequently, v is monotone. From the difference equation we find u_{j+1}:

$$u_{j+1} = (1 - \tau a/b)^{j+1}.$$

This means that for the function u to be monotone it is necessary that $b - \tau a \geq 0$. Otherwise, the solution will be oscillating. Thus, if we pass to ill-posed problems, the classical stability condition of Samarskii (1977) $B \geq \tau A/2$ is replaced by the stronger condition $B \geq \tau A$. As it follows from the example considered above, this replacement is quite natural.

Now, we consider the perturbation of the explicit scheme $Pu = u_t - Au$ by the variable operator K:

$$Qu = u_t - Au - Ku = f. \qquad (3.10)$$

We show that the stability is conserved if the operator K, in a certain sense, is subordinate to the operator A.

Theorem 3.6. Suppose $A = A^*$, $K \in \mathcal{L}(H)$,

$$E + \tau A \geq 0 \qquad (3.11)$$

and for all $u \in H$

$$\|Ku\|^2 \leq c(|\langle Au, u\rangle| + \|u\|^2). \qquad (3.12)$$

Then there exists a number c_1 which depends only on $c \geq 0$, such that the scheme (3.10) is finitely stable under the condition

$$\varphi_{t\bar{t}} + c_1\varphi_{\bar{t}} - c_1 \geq 1. \qquad (3.13)$$

To prove this theorem, we need the following statement.

Lemma 3.1. Suppose that for all $s \geq s_0$, $v \in C_0(Z_0^N)$ and for some constants c_0 and c the following estimate holds:

$$[s\varphi_{t\bar{t}}, \|v\|^2] + c_0\|Av + s\varphi_t\hat{v}\|_0^2 \leq c\|Pv - s\varphi_t\hat{v}\|_0^2, \qquad (3.14)$$

and the operator K satisfies estimate (3.12). Then, under condition (3.13), where $c_1 = c_1(c_0, c)$, the difference scheme (3.10) is finitely stable. Here $\|v\|_0^2 = \tau\sum_{j=0}^{N-1}\|v_j\|^2$.

Proof. Since $(a+b)^2 \leq 2a^2 + 2b^2$, we have

$$\|Pv - s\varphi_t\hat{v}\|_0^2 \leq 2\|Qv - s\varphi_t\hat{v}\|_0^2 + 2\|Kv\|_0^2. \tag{3.15}$$

On the other hand, using the estimate

$$2|ab| \leq \alpha a^2 + \alpha^{-1}b^2,$$

from (3.12) we obtain

$$\|Kv\|^2 \leq c(|\langle Av + s\varphi_t\hat{v} - s\varphi_t\hat{v}, v\rangle| + \|v\|^2)$$
$$\leq c|\langle Av + s\varphi_t\hat{v}, v\rangle| - cs\varphi_t|\langle \hat{v}, v\rangle| + c\|v\|^2$$
$$\leq c\alpha\|Av + s\varphi_t\hat{v}\|^2/2 - s\varphi_t c\|\hat{v}\|^2/2 + \|v\|^2(-s\varphi_t c/2 + c + c/2\alpha).$$

Therefore, due to the inequality $[\varphi_t, \|\hat{v}\|^2] = (\varphi_{\bar{t}}, \|v\|^2)$, we have

$$2\|Kv\|_0^2 \leq c\alpha\|Av + s\varphi_t\hat{v}\|_0^2 + (-sc(\varphi_t + \varphi_{\bar{t}}) + c/\alpha + 2c, \|v\|^2).$$

Combining this estimate with inequalities (3.14) and (3.15), we find

$$s(\varphi_{t\bar{t}} + c^2\varphi_t + c^2\varphi_{\bar{t}} - c(c/\alpha + 2c)/s, \|v\|^2) + (c_0 - c^2\alpha)\|Av + s\varphi_t\hat{v}\|_0^2$$
$$\leq 2c\|Qv - s\varphi_t\hat{v}\|_0^2.$$

Choosing α so that $c_0 - c^2\alpha \geq 0$ and setting $c_1 = \max(2c^2, c^2(1/\alpha + 2))$, for $s \geq 1$ we have

$$s(1, \|v\|^2) \leq 2c\|Qv - s\varphi_t\hat{v}\|_0^2.$$

Hence, for $v = u\Psi$ we obtain the estimate

$$s\|u\|_s^2 \leq 2c\|Qu\|_s^2,$$

which is the required conclusion. $\square$

To prove Theorem 3.6, it remains to note that under the condition $E + \tau A \geq 0$, by Theorems 2.1, 3.1, we have estimate (2.23). This estimate is transformed into estimate (3.14) of Lemma 3.1 for $v = \Psi u$.

Now, we consider the scheme with weights containing the skew-symmetric term iBu:

$$Pu = u_t - (A + iB)(\sigma\hat{u} + (1-\sigma)u). \tag{3.16}$$

We also assume that the self-adjoint operators A and B commute and are constant.

Chapter 6. Theory of stability of difference schemes

Theorem 3.7. *Suppose that for some $\delta, m, c > 0$*

$$E - \sigma\tau A \geq \delta E, \qquad (3.17)$$

$$(1 - 2\sigma)A \geq -mE, \qquad (3.18)$$

$$|1 - \sigma|\tau\|B\|^2 \leq c. \qquad (3.19)$$

Then there exists a number $c_1 = c_1(m, \delta, c)$ such that the scheme (3.16) is stable for

$$\varphi_{t\bar{t}} + c_1\varphi_t \geq 1. \qquad (3.20)$$

Proof. In this case, the operators M_1 and M_0 from Theorem 2.1 are as follows:

$$M_1 = s\varphi_{t\bar{t}}(E - \sigma\tau A) + |1 - \sigma|\tau\varphi_t s\alpha\|B\|^2 E,$$
$$M_0 = \tau\{-s\varphi_t(E - \sigma\tau A) + |1 - \sigma|s\varphi_t\alpha^{-1} + (1 - 2\sigma)A\}.$$

Choosing the numbers s_0 and α sufficiently large and taking into account conditions (3.1), (3.17), and (3.18), we see that the operator M_0 is nonnegative. Analogously, conditions (3.17), (3.19), and (3.20) yield $M_1 \geq s\varepsilon E$ for sufficiently large c_1 and sufficiently small $\varepsilon > 0$. It remains to apply Theorem 2.1. □

We now consider the case of variable operators A and B setting $\sigma \in [0, 1]$. First, we assume that $B \equiv 0$.

Theorem 3.8. *Suppose $A = A^*$, $-A_t \leq cA$ for some number $c \geq 0$. If*

$$E - \tau\sigma A \geq \delta E, \qquad (3.21)$$

$$\sigma m E + \sigma c(1 - \sigma)\tau A \geq 0, \qquad (3.22)$$

$$E + \tau(1 - 2\sigma)A \geq c\tau E, \qquad (3.23)$$

$$mE - c(1 - \sigma)^2\tau A \geq 0, \qquad (3.24)$$

for some numbers $\delta, m > 0$, then there exist numbers $c_0, \tau_0 > 0$ such that the scheme with weights, $Pu = u_t - A(\sigma\hat{u} + (1 - \sigma)u)$, is finitely stable for

$$\varphi_{t\bar{t}} + c_0\varphi_{\bar{t}} - c_0 \geq 1. \qquad \tau \leq \tau_0 \qquad (3.25)$$

Proof. Inequalities (2.25) and (2.18) yield

$$N_0 = (E + \tau\lambda(1-2\sigma)A) - \tau s\varphi_t(E - \lambda\sigma\tau A) - (\tau\lambda c\tau(1-\sigma)^2/(1-\sigma s\varphi_t\tau))A.$$

By the definition of the function λ (see (2.17)), we have

$$\lambda = (1 - \tau\chi)^{-1},$$

$$c/2 \leq \chi = c(1 + (1-\sigma)/(1 - \sigma s\tau\varphi_t))/2 \leq c, \tag{3.26}$$

$$\lambda^{-1} = 1 - \tau\chi. \tag{3.27}$$

From formulas (3.23), (3.26), and (3.27) it follows that the first term in N_0 is nonnegative. Indeed,

$$E + \tau\lambda(1-2\sigma)A = \lambda(E + \tau(1-2\sigma)A - \tau\chi E)$$
$$\geq \lambda(E + \tau(1-2\sigma)A - c\tau E) \geq 0.$$

The sum of two last terms in N_0 is greater or equal to

$$-\tau s\varphi_t\delta/2\{E - 2\lambda\tau c(1-\sigma)^2/(-s\varphi_t\delta(1-s\sigma\tau\varphi_t))A\}. \tag{3.28}$$

Here we have used that

$$E - \lambda\sigma\tau A = \lambda(E - \sigma\tau A - \tau\chi E) \geq \delta/2\, E \tag{3.29}$$

if the condition (3.21) holds and τ_0 is sufficiently small. In its turn, the expression in braces in (3.28) is equal to

$$\frac{2\lambda}{-s\varphi_t(1-\sigma s\tau\varphi_t)\delta}\left\{-s\varphi_t\frac{\delta}{2}(1-\sigma s\tau\varphi_t)(1-\tau\chi) - c(1-\sigma)^2\tau A\right\}.$$

Therefore, for $s \geq s_0$ it is nonnegative (if s_0 is sufficiently large) due to condition (3.24) and the relation $-\varphi_t \geq 1$. Thus, we have proved that $N_0 \geq 0$. In view of (2.24), (2.16), and (2.17) we have

$$N_1 = s\varphi_{t\bar{t}}(E - \lambda\sigma\tau A) + \lambda_t(1 - \sigma s\tau\varphi_t)A + \{\bar{\lambda}cs(\varphi_t + \varphi_{\bar{t}})/2 - \bar{\lambda}c/2\beta\}E, \tag{3.30}$$

$$\lambda_t = \frac{\lambda\hat{\lambda}c(1-\sigma)\sigma s\tau^2\varphi_{tt}}{2(1-\sigma s\tau\hat{\varphi}_t)(1-\sigma s\tau\varphi_t)}$$

and therefore the sum of the second term and the half of the first one is greater or equal to

$$N_2 \equiv \frac{s\varphi_{t\bar{t}}\delta}{4}E + \lambda_t(1 - \sigma s\tau\varphi_t)A$$

$$= \frac{\lambda\hat{\lambda}s\varphi_{tt}}{1-\sigma s\tau\hat{\varphi}_t}\left\{\frac{s\varphi_{t\bar{t}}\delta}{4}\frac{1-\sigma s\tau\hat{\varphi}_t}{\lambda\hat{\lambda}s\varphi_{tt}}E + \frac{c}{2}\sigma(1-\sigma)\tau^2 A\right\}.$$

Chapter 6. Theory of stability of difference schemes

Since
$$\frac{1 - \sigma s \tau \hat{\varphi}_t}{s} \geq -\sigma \tau \hat{\varphi}_t,$$

the operator N_2 satisfies the inequality

$$N_2 \geq \frac{\lambda \hat{\lambda} s \varphi_t t}{1 - \sigma s \tau \hat{\varphi}_t} \sigma \tau \left\{ \frac{s \varphi_{t\bar{t}} \delta}{4} \frac{-\hat{\varphi}_t E}{\lambda \hat{\lambda} \varphi_{tt}} + \frac{c}{2}(1-\sigma)\tau A \right\} \geq 0$$

for $s \geq s_0$ and sufficiently large s_0. We have also used estimate (3.22). From condition (3.25) and the relation $-\varphi_{\bar{t}} \geq -\varphi_t$ it follows that the sum of the remaining half of the first term and the last term of (3.30) is not less than $s\varphi_{t\bar{t}} \delta E/4$ if c_0 is sufficiently large. Therefore,

$$N_1 \geq s\varphi_{t\bar{t}} \delta/4 \geq s\delta/4 \, E,$$

and the assertion of the theorem follows from Theorem 2.1. □

We now consider the case $B \neq 0$. From (2.21) it follows that two new terms Δ appear in the operator M_1:

$$\Delta = (1-\sigma)\tau s \varphi_t \alpha \|B\|^2 E - L. \tag{3.31}$$

Here, by formula (2.19) we have

$$-L = c_1(s(\varphi_t + \varphi_{\bar{t}})/2 - 1/2\gamma).$$

If we assume that $(1-\sigma)\tau\|B\|^2 \leq m$, then the first term in (3.31) is estimated from below by the quantity $s\varphi_t \alpha m$. According to the proof of Theorem 3.8, we have $N_1 \geq s\delta/4 \, \varphi_{t\bar{t}} E$ and therefore

$$M_1 = N_1 + \Delta \geq s\varepsilon \varphi_{tt} E \geq s\varepsilon E$$

for some positive $\varepsilon < \delta/4$ if the function φ satisfies condition (3.25) with sufficiently large constant c_0. We find the sufficient conditions of nonnegativeness

$$M_0 = -\tau s \varphi_t (E - \tau \sigma \lambda A) + (1-\sigma)\tau s \alpha^{-1} \varphi_t E$$
$$- c\lambda \tau^2 (1-\sigma)^2/(1 - \sigma s \tau \varphi_t) A + \tau \lambda (1-2\sigma) A.$$

For $(1-\sigma)\alpha^{-1} = \delta/4$, (3.29) implies

$$M_0 \geq -\frac{\tau s \varphi_t \delta}{4} - \frac{c\lambda \tau^2 (1-\sigma)^2}{1 - \sigma s \tau \varphi_t} A + \tau \lambda (1-2\sigma) A$$
$$= \frac{\lambda \tau}{1 - \sigma s \tau \varphi_t} \left\{ -\frac{s \varphi_t \delta (1 - \sigma s \tau \varphi_t) E}{4 \lambda} - c\tau(1-\sigma)^2 A \right\} + \tau \lambda(1-2\sigma) A \geq 0,$$

if $(1-2\sigma)A \geq 0$, $mE - c(1-\sigma)^2\tau A \geq 0$, $s \geq s_0$, and the number s_0 is sufficiently large. Due to Theorem 2.1 these estimates imply the following statement.

Theorem 3.9. *Suppose* $A = A^*$, $B = B^*$, $-A_t \leq cA$, $-[A, iB] \leq c_1 A$ *for some numbers* $c, c_1 \geq 0$, *where* $c_1 = 0$ *for* $\sigma > 0$. *If for some numbers* $\delta, m > 0$ *we have*

$$E - \tau\sigma A \geq \delta E, \tag{3.32}$$

$$\delta mE + \sigma c(1-\sigma)\tau A \geq 0, \tag{3.33}$$

$$mE - c(1-\sigma)^2\tau A \geq 0, \tag{3.34}$$

$$(1-2\sigma)A \geq 0, \tag{3.35}$$

$$(1-\sigma)\tau\|B\|^2 \leq m,$$

then there exist numbers $c_0, \tau_0 > 0$ *such that the scheme with weights*

$$Pu = u_t - (A + iB)(\sigma\hat{u} + (1-\sigma)u)$$

is finitely stable under the conditions

$$-\varphi_t \geq 1, \qquad \varphi_{t\bar{t}} + c_0\varphi_{\bar{t}} - c_0 \geq 1, \qquad \tau \leq \tau_0$$

In typical ill-posed cases, $A \geq 0$. Under this condition, the following theorem holds.

Theorem 3.10. *Suppose* $0 \leq \sigma \leq 1/2$ *and the following conditions hold for all* $j \in Z_0^{N-1}$:

$$A = A^* \geq 0, \qquad B = B^*,$$

$$-A_t \leq cA, \qquad -i[A, B] \leq c_1 A,$$

$$\sigma\tau\|A\| \leq q < 1, \qquad (1-\sigma)\tau\|B\|^2 \leq c,$$

$$-\varphi_t \geq 1, \qquad \varphi_{t\bar{t}} + c_0\varphi_{\bar{t}} - c_0 \geq 1$$

Here $c, c_1 \geq 0$, $c_1 = 0$ *for* $\sigma > 0$ *and* c_0 *is sufficiently large and depends on* c, c_1, *and* q. *Then there exist positive numbers* s_0 *and* τ_0 *such that for all* $s \geq s_0$, $u \in C_0(Z_0^N)$ *the following estimate holds for* $\tau \geq \tau_0$:

$$s\|u\|_s^2 \leq \text{const}\,\|Pu\|_s^2.$$

Proof. Since $\sigma\tau\|A\| \leq q < 1$, $A \geq 0$, and $\sigma \leq 1/2$, conditions (3.32), (3.33), and (3.35) of Theorem (3.9) hold automatically. Therefore, $M_1 \geq s\varepsilon E$ for some $\varepsilon > 0$. In the operator

$$M_0 = -\tau s\varphi_t(E - \tau\lambda\sigma A) + (1-\sigma)\tau s\alpha^{-1}\varphi_t E$$
$$+ \tau\lambda\{(1-2\sigma) - c\tau(1-\sigma)^2/(1-\sigma s\tau\varphi_t)\}A$$

the first two terms are estimated from below, as in Theorem 3.9, by the quantity $-\tau\varphi_t \cdot \delta/4\, E \geq 0$. The nonnegativeness of the third term follows from $A \geq 0$ and the multiplier in braces is nonnegative if $\sigma < 1/2$ and the number $\tau \leq \tau_0$, where τ_0 is sufficiently small. If $\sigma = 1/2$, then condition (3.34) of Theorem (3.9) holds. Therefore, we have $M_0 \geq 0$ also in this case. The theorem is proved. □

The stability of the explicit difference scheme is conserved also for the perturbations described in the theorem below.

Theorem 3.11. *Suppose that the assumptions of one of Theorems 3.7–3.10 hold with $\sigma = 0$. If the operator $K \in \mathcal{L}(H)$ satisfies the estimate*

$$\|Ku\|^2 \leq c(|\langle Au, u\rangle| + \|u\|^2), \qquad \forall u \in H,$$

then the explicit scheme $Qu = Pu + Ku$ is finitely stable.

Theorem 3.11 is proved using Lemma 3.1 similarly to the proof of Theorem 3.6.

Exercise 3.1. Suppose $Pu = u_t - Au$, $A \in \mathcal{L}(H)$, $A^* = A$, and

$$(1 + s_0\tau)E + \tau A \geq 0$$

for some $s_0 \geq 0$. Assume also that φ satisfies the condition

$$\varphi_{t\bar{t}} \geq 1, \qquad -\varphi_t \geq 1.$$

Prove that the estimate

$$s\|u\|_s^2 \leq \|Pu\|_s^2$$

holds for all $s \geq s_0$ and $u \in C_0(Z_0^N)$.

Exercise 3.2. Explain why the implicit scheme $\Psi_t = s\varphi_t\Psi$ is chosen for determination of the weight function Ψ. What shall we get if we use the explicit scheme $\Psi_t = s\varphi_t\Psi$, $\Psi_0 = 1$ to determine the function Ψ?

Exercise 3.3. Consider the difference scheme with operator weight $\sigma \in \mathcal{L}(H)$:
$$Pu = u_t - \sigma A\hat{u} - (E - \sigma)Au.$$
Assuming $A = A^*$, $\sigma = \sigma^*$, and $\sigma A = A\sigma$, prove that if
$$E - \tau\sigma A \geq \delta E, \qquad \delta > 0, \quad E + \tau(E - 2\sigma)A \geq 0,$$
then the scheme P is finitely stable.

Exercise 3.4. Prove formulas (2.6), (2.8).

Exercise 3.5. Suppose that a linear operator $A(t) : V \to H$ is defined for any $t \in [0, T]$, where V is a linear manifold in a Hilbert space H which is independent of t. We consider the Cauchy problem for the equation of the first order
$$Pu = du/dt - A(t)u = f(t),$$
$$u(0) = g \in V. \qquad (3.36)$$
For the domain of definition of the operator P we take the linear manifold
$$D(P) = \{u \in C^1([0, T]; H) \mid u, u'(t) \in V\},$$
assuming that the operator A is symmetric $\langle Au, u \rangle = \langle u, Au \rangle$, strongly continuously differentiable with respect to t in $D(A) = V$ and
$$-\langle A'_t u, u \rangle \leq c(|\langle Au, u \rangle| + \|u\|^2), \qquad \forall u \in V.$$
Here, by $C^1([0, T]; H)$ we denote the Banach space of continuously differentiable functions $u : [0, T] \to H$ with values in H
$$\|u\|_{C^1([0,T];H)} = \sup_{t \in [0,T]} \|u(t)\| + \sup_{t \in [0,T]} \|u'(t)\|.$$
We set
$$D_0(P) = D(P) \cap \{u \in C^1([0, T]; H) \mid u(0) = u(T) = 0\},$$
$$\varphi(t) = \exp(\mu(T - t)), \qquad \mu = 2 + c + \lambda,$$
$$s_0 = (c + \lambda)^2/12, \qquad \lambda \geq 0.$$

Prove that the following estimate holds for all $s \geq s_0$, $u \in D_0(P)$:

$$s \int_0^T \exp(2s\varphi(t)) \|u(t)\|^2 \, dt + \lambda \int_0^T \exp(2s\varphi(t)) \, | \, \langle A(t)u(t), u(t)\rangle | \, dt$$

$$\leq \int_0^T \exp(2s\varphi(t)) \|Pu\|^2 \, dt. \tag{3.37}$$

(Directions: follow the proofs of Theorems 3.7, 3.11, passing to the limit as $\tau \to 0$.)

Exercise 3.6. Taking estimate (3.37) into account, prove the uniqueness for the problem (3.36) and establish the stability estimate

$$\|u\|_{L_2(0,T';H)} \leq \varepsilon \|u\|_{L_2(T',T;H)}$$
$$+ c(\varepsilon, T', T)[\|Pu\|_{L_2(0,T;H)} + \|A(t)g\|_{L_2(0,T;H)} + \|g\|],$$

where $T' < T$.

Exercise 3.7. Suppose that $u(x,t)$ is a solution of the boundary value problem

$$\frac{\partial u}{\partial t} + \sum_{i,j=1}^n a_{ij}(x,t) \frac{\partial^2 u}{\partial x_i \partial x_j} + \sum_{k=1}^n a_k(x,t) \frac{\partial u}{\partial x_k} + a_0(x,t)u = f(x,t),$$

$$u(x,0) = 0, \qquad x \in \Omega; \qquad u|_{\partial\Omega \times [0,T]} = 0, \tag{3.38}$$

where Ω is a bounded domain in $\mathbb{R}^n$ with smooth boundary, the coefficients a_{ij} and a_k are sufficiently smooth in $\Omega \times [0,T]$ and

$$c_1|\xi|^2 \geq \sum_{i,j=1}^n a_{ij}(x,t)\xi_i\xi_j \geq c_0|\xi|^2.$$

Using Exercises 3.5, 3.6, establish the uniqueness and l-stability for this boundary value problem.

Exercise 3.8. Consider Exercise 3.7 replacing condition (3.38) by the condition

$$\left[\frac{\partial u}{\partial \nu} - \sigma(x,t)u\right]\bigg|_{\partial\Omega \times [0,T]} = 0.$$

Here, ν is the exterior normal to $\partial\Omega$ and $\sigma(x,t)$ is a smooth function defined in $\partial\Omega \times [0,T]$.

6.4. ESTIMATES OF l-STABILITY UP TO THE BOUNDARY

In Section 1, it was shown that finite stability of a difference scheme implies its l-stability in the norm $l_2(1, N''; H)$. Here $N'' \le N' < N$ and the stabilizing functional l contains the quantities u_n, $n = N'', \ldots, N$. Thus, the stability of the problem was guaranteed only in the "interior domain" of the mesh Z_1^N. To obtain the stability estimate in the entire mesh Z_1^N, we should take into account the contribution of nonintegral terms appearing after integrating by parts. Therefore, we should deal not with arbitrary functions $u : Z_0^N \to H$ rather than with finite functions $C_0(Z_0^N)$. For simplicity, we shall perform calculations restricting ourselves to the case of the following scheme with weights:

$$Pu = u_t - A(\sigma \hat{u} + (1-\sigma)u) = f, \qquad u_0 = g, \qquad (4.1)$$

where the operator A is assumed to be self-adjoint and independent of j.

To obtain the stability estimate we shall estimate $\|Pu\|_s^2$ from below for the explicit scheme, $\sigma = 0$. We have

$$\|Pu\|_s^2 = \tau \sum_{j=0}^{N-1} \|u_t - Au\|^2 \Psi_j^2(s).$$

As in Section 2, making the change of variables

$$u = v/\Psi, \qquad u_t = (v_t - s\varphi_t \hat{v})/\Psi,$$

we obtain

$$\|Pu\|_s^2 = \tau \sum_{j=0}^{N-1} \|v_t - s\varphi_t \hat{v} - Av\|^2$$

$$= \tau \sum_{j=1}^{N-1} \{\|v_t\|^2 + \|Av + s\varphi_t \hat{v}\|^2 - 2\operatorname{Re}\langle v_t, s\varphi_t \hat{v}\rangle - 2\operatorname{Re}\langle v_t, Av\rangle\}$$

$$= I_1 + I_2 + I_3 + I_4. \qquad (4.2)$$

Taking into account the designations (2.4) and (2.5), we have

$$I_1 = [1, \|v_t\|^2] \ge 0, \qquad I_2 = [1, \|Av + s\varphi_t \hat{v}\|^2] \ge 0,$$
$$I_3 = -[s\varphi_t, 2\operatorname{Re}\langle v_t, \hat{v}\rangle], \qquad I_4 = -[1, 2\operatorname{Re}\langle v_t, Av\rangle]. \qquad (4.3)$$

Chapter 6. Theory of stability of difference schemes

We transform the values I_3 and I_4 by means of integrating by parts

$$[x, \partial y] = -(\bar{\partial} x, y) + x_{N-1} y_N - x_0 y_0. \tag{4.4}$$

Using identity (2.7), we obtain

$$I_3 = [-s\varphi_t, \partial \|v\|^2) - [s\tau\varphi_t, \|v_t\|^2)$$
$$= (s\varphi_{t\bar{t}}, \|v\|^2) - [s\tau\varphi_t, \|v_t\|^2) - s((\varphi_t)_{N-1}\|v_N\|^2 - (\varphi_t)_0\|v_0\|^2). \tag{4.5}$$

We assume now that

$$m \geq -\varphi_t \geq \mu > 0, \qquad \varphi_{t\bar{t}} \geq \mu_1 > 0.$$

These inequalities and (4.5) yield

$$I_3 \geq s\mu_1 \|v\|^2_{l_2(1,N-1;H)} + s\mu\|v_N\|^2 - sm\|v_0\|^2. \tag{4.6}$$

Since $2\,\mathrm{Re}\langle v_t, Av\rangle = \partial(v, Av) - \tau\langle v_t, Av_t\rangle$, using formulas (4.4), we have

$$I_4 = -[1, \partial\langle v, Av\rangle) + [1, \langle v_t, \tau Av_t\rangle) = [1, \langle v_t, \tau Av_t\rangle) - \langle v_N, Av_N\rangle + \langle v_0, Av_0\rangle.$$

Summing up I_1, I_3, and I_4 and omitting $I_2 \geq 0$, by formulas (4.2), (4.3), (4.5), and (4.6), we have

$$\|Pu\|^2_s \geq s\mu_1\|v\|^2_{l(1,N-1;H)} + [1, \langle v_t, (E+\tau A)v_t\rangle)$$
$$+ s\mu\|v_N\|^2 - sm\|v_0\|^2 - \langle v_N, Av_N\rangle + \langle v_0, Av_0\rangle. \tag{4.7}$$

Any self-adjoint operator $A \in \mathcal{L}(H)$ can be represented as the difference of two commuting self-adjoint nonnegative operators $A_\pm$:

$$A = A_+ - A_-, \qquad A_+ \geq 0, \quad A_- \geq 0. \tag{4.8}$$

So, assuming in addition that

$$E + \tau A \geq 0, \tag{4.9}$$

from (4.8) and (4.9) we obtain the following rough version of inequality (4.7):

$$\|Pu\|^2_s \geq s\mu_1\|v\|^2_{l_2(1,N-1;H)} + s\mu\|v_N\|^2 - sm\|v_0\|^2$$
$$- \langle v_n, A_+ v_N\rangle - \langle v_0, A_- v_0\rangle. \tag{4.10}$$

For $0 < \tau \leq \tau_0$ and $\mu_2 = \min(\mu_1, \mu/\tau_0)$, we have the estimate

$$s\mu_1\|v\|^2_{l_2(1,N-1;H)} + s\mu\|v_N\|^2 \geq s\mu_2\|v\|^2_{l_2(1,N;H)}. \tag{4.11}$$

Recalling that $v = \Psi u$ and $\Psi_{j+1} \leq \Psi_j$, from (4.10) and (4.11) we obtain

$$\|Pu\|_{l_2(0,N-1;H)}^2 \geq \|Pu\|_s^2 \geq s\mu_2 \Psi_N^2 \|u\|_{l_2(1,N;H)}^2 - sm\|u_0\|^2$$
$$- \langle u_0, A_- u_0 \rangle - \Psi_N^2 \langle u_N, A_+ u_N \rangle,$$

which implies

$$s\mu_2 \Psi_N^2 \|u\|_{l_2(1,N;H)}^2 \leq \Psi_N^2 \|A_+^{1/2} u_N\|^2 + \|Pu\|_{l_2(0,N-1;H)}^2$$
$$+ sm\|u_0\|^2 + \|A_-^{1/2} u_0\|^2.$$

Divide both sides of this inequality by $s\mu_2 \Psi_N^2$ and set

$$\varepsilon^2 = 1/s\mu_2,$$
$$c^2(\varepsilon) = \max((s\mu_2)^{-1} \exp(2smT), m\mu_2^{-1} \exp(2smT)),$$
$$\|u_0\|_{A_-}^2 = \|u_0\|^2 + \|A_-^{1/2} u_0\|^2.$$

Then, taking into account the inequality $1/\Psi_N \leq \exp(smT)$ and Lemma 1.1, we obtain

$$\|u\|_{l_2(1,N;H)}^2 \leq \varepsilon^2 \|A_+^{1/2} u_N\|^2 + c^2(\varepsilon)[\|Pu\|_{l_2(0,N-1;H)}^2 + \|u_0\|_{A_-}^2]. \quad (4.12)$$

Thus, for

$$l^2(u) = \|A_+^{1/2} u_N\|^2, \qquad D(l) = l_2(1, N; H)$$

we see that the explicit difference scheme

$$Pu = u_t - Au, \qquad u_0 = g$$

is l-stable with respect to the right-hand side and the initial data g. In other words, we have proved the following statement.

Theorem 4.1. *Suppose $A = A^* = A_+ - A_-$, where $(A_\pm)^* = A_\pm \geq 0$, and the condition $E+\tau A \geq 0$ holds. Then the stability estimate (4.12) holds for the explicit scheme $Pu = u_t - Au$ for any $\tau \in (0, \tau_0]$, $\varepsilon > 0$, $u: Z_0^N \to H$.*

Note that the function $c(\varepsilon)$ from estimate (4.12) is independent of $\tau \in (0, \tau_0]$ and $\|A\|$. It is necessary only that concordance condition (4.9) holds. In particular, it holds if $\tau \|A_-\| \leq 1$.

Now, we transform the scheme with weights (4.1) into the explicit scheme with $\sigma = 0$. To this end, using the identity

$$\sigma \hat{u} = \sigma u + \tau \sigma u_t,$$

Chapter 6. Theory of stability of difference schemes

we rewrite the scheme (4.1) in the form

$$Pu = (E - \sigma\tau A)u_t - Au = f.$$

Assuming that

$$(E - \sigma\tau A)^{-1} \in \mathcal{L}(H), \qquad (4.13)$$

we have

$$u_t - \overline{A}u = \overline{f},$$

where $\overline{A} = (E - \sigma\tau A)^{-1}A$ and $\overline{f} = (E - \sigma\tau A)^{-1}f$.
Condition (4.13) will hold if

$$E - \tau\sigma A \geq \delta E, \qquad \delta > 0. \qquad (4.14)$$

In this case,

$$\|\overline{f}\| \leq \delta^{-1}\|f\|,$$

$$\overline{A}_\pm = (E - \sigma\tau A)^{-1}A_\pm, \qquad (4.15)$$

and the condition $E + \tau\overline{A} \geq 0$ is equivalent to the inequality

$$E + \tau(1 - \sigma)A \geq 0. \qquad (4.16)$$

Thus, Theorem 4.1 implies the following assertion:

Theorem 4.2. *Suppose $A = A^*$ and conditions (4.14) and (4.16) hold. Then, for any $\tau \in (0, \tau_0]$, $\varepsilon > 0$ and $u : Z_0^N \to H$ the following stability estimate holds*

$$\|u\|_{l_1(1,N;H)}^2 \leq \varepsilon^2 \|\overline{A}_+^{1/2} u_N\|^2 + c^2(\varepsilon)[\delta^{-2}\|Pu\|_{l_2(0,N-1;H)}^2 + \|u_0\|_{\overline{A}_-}^2],$$

where the operators $\overline{A}_\pm$ are defined by formula (4.15).

Remark 4.1. From the proof of Theorems 4.1 and 4.2 it follows that

$$c(\varepsilon) = O(\exp(c_0/\varepsilon^2)) \qquad (4.17)$$

for $\varepsilon \to 0$, where c_0 depends only on m, μ, μ_2, τ_0, and T.

Exercise 4.1. Generalize Theorems 4.1 and 4.2 to the case of a variable self-adjoint operator A such that $-A_t \leq cA$.

Exercise 4.2. By means of passing to the limit, prove the analogue of Theorem 4.1 for the abstract Cauchy problem

$$\frac{du}{dt} - Au = f(t), \qquad t \in [0,T], \quad u(0) = g,$$

where $A : V \to H$, V is a linear manifold which is dense in H, and the linear operator A is self-adjoint and independent of t.

6.5. CONVERGENCE THEOREMS

In Chapter 5, it was shown that the solution of the l_h-well-posed difference scheme converges to the solution of the exact problem if the conditions of approximation, of the *a priori* smoothness of the solution, and of concordance of the scheme parameters with the accuracy of initial data are valid.

In this section Theorem 2.1 of Chapter 5 is refined to be applied to the abstract Cauchy problem.

Suppose v is a solution of the following problem

$$dv/dt = Av, \tag{5.1}$$

$$v(0) = v_0. \tag{5.2}$$

Here A is a linear unbounded operator independent of t with domain of definition $D(A) \subseteq H$. We suppose that H is a Hilbert space, the function $v : [0,T] \to H$ is twice strongly continuously differentiable, i.e., $v \in C^2([0,T]; H)$, where $v(t) \in D(A)$ for any $t \in [0,T]$. We consider the following one-parameter set of difference schemes corresponding to the problem (5.1), (5.2):

$$u_t - A_h(\sigma\hat{u} + (1-\sigma)u) = 0, \qquad \sigma \in \mathbb{R}, \tag{5.3}$$

$$u_0 = g. \tag{5.4}$$

We assume that $A_h \in \mathcal{L}(H_h)$, $A_h = A_h^* \geq 0$, and the operator A_h approximates the operator A in a certain sense which is specified below. In its turn, the Hilbert space H_h, which depends on the parameter $h > 0$, approximates the space H. To specify in detail the notion of approximation, we assume that there exists a set of linear operators

$$P_h : V \to H_h$$

Chapter 6. Theory of stability of difference schemes

defined in a linear manifold $V \supseteq D(A) \cup R(A)$ dense in H, where $R(A) = AD(A)$ is the range of values of the operator A, such that

$$\lim_{h \to 0} \|P_h u\| = \|v\|_H, \qquad \forall v \in V. \tag{5.5}$$

Hereinafter $\|\cdot\|$ is a norm in H_h.

Consider the problem of convergence of u to v in the following interpretation

$$\|u - P_h v\|_{l_2(1,N;H_h)} \to 0$$

for $h \to 0$, $\tau \to 0$.

Theorem 5.1. *Suppose that the following conditions hold:*

1. *The conditions of approximation of the operator A and the initial data v_0:*

$$\|(A_h P_h - P_h A)v\| \leq c_1 h^p, \tag{5.6}$$

$$\|g - P_h v_0\| \leq \delta. \tag{5.7}$$

2. *The stability condition*

$$\sigma\tau\|A_h\| \leq 1 - q, \qquad q \in (0, 1).$$

3. *The smoothness conditions for the solution v of the problem (5.1), (5.2):*

$$\|A_h P_h v_t\| \leq c_1, \qquad \|P_h v''\| \leq c_1, \tag{5.8}$$

$$\|\overline{A}_h^{1/2} P_h v_N\| \leq m, \qquad v_N = v(N\tau) = v(T), \tag{5.9}$$

$$\|\overline{A}_h^{1/2} S^N P_h v_0\| \leq m, \tag{5.10}$$

where $S = E + \tau \overline{A}_h$, $\overline{A}_h = (E - \sigma\tau A_h)^{-1} A_h$.

4. *The concordance condition for the parameters h, δ, and σ:*

$$\|\overline{A}_h\|^{1/2} \exp(T\|\overline{A}_h\|)\delta \leq c_2. \tag{5.11}$$

Then the following convergence estimate holds:

$$\|u - P_h v\|_{l_2(1,N;H)} \leq \omega_{2m+c_2}(\tau + h^p + \delta) \to 0, \qquad \delta, \tau, h \to 0$$

where $\omega_m(\delta) = O(1/\ln \delta^{-1})^{1/2}$.

Proof. We set $w = u - P_h v$,

$$u = w + P_h v. \tag{5.12}$$

Substituting (5.12) into equation (5.3), we obtain

$$w_t - A_h(\sigma \hat{w} + (1-\sigma)w) = f, \tag{5.13}$$

$$w_0 = g - P_h v_0, \tag{5.14}$$

where

$$f = A_h(\sigma P_h \hat{v} + (1-\sigma) P_h v) - P_h v_t.$$

Estimate the norm of f. By the identity

$$\sigma P_h \hat{v} + (1-\sigma) P_h v = \sigma \tau P_h v_t + P_h v$$

we have

$$f = \sigma \tau A_h P_h v_t + A_h P_h v - P_h v_t. \tag{5.15}$$

Since $v \in C^2([0,T];H)$ by the assumption, we have

$$(v_t)_j = \frac{v(j\tau + \tau) - v(j\tau)}{\tau} = v'(j\tau) + \frac{\tau}{2} v''(j\tau + \theta\tau),$$

$$\theta \in (0,1), \qquad j = 0, \ldots, N-1, \qquad N\tau = T.$$

Hence, by the second estimate of (5.8),

$$P_h v_t = P_h v'(j\tau) + O(\tau). \tag{5.16}$$

More precisely,

$$\|P_h v_t - P_h v'(j\tau)\| \leq (\tau/2) c_1.$$

Analogously, the condition of approximation, (5.6), leads to the equality

$$A_h P_h v = P_h A v + O(h^p). \tag{5.17}$$

As a result, from (5.15)–(5.17) and $P_h(v' - Av) = 0$, in view of (5.1) we have

$$f = \sigma \tau A_h P_h v_t + O(\tau + h^p).$$

Taking into account the first condition of (5.8), we obtain

$$\|f_j\|_{H_h} \leq |\sigma| \tau c_1 + O(\tau + h^p),$$

Chapter 6. Theory of stability of difference schemes

$$\|f\|_{l_2(0,N-1;H)} = \left\{\tau \sum_{j=0}^{N-1} \|f_j\|_{H_h}^2\right\}^{1/2} \le |\sigma|\tau c_1 \sqrt{T} + O(\tau + h^p). \quad (5.18)$$

Now, we estimate $\|\overline{A}_h^{1/2} w_N\|_{H_h}$. As far as $w_N = u_N - P_h v_N$ and $v_N = v(T)$, by the triangle inequality and condition (5.9)

$$\|\overline{A}_h^{1/2} w_N\| \le \|\overline{A}_h^{1/2} u_N\| + m. \quad (5.19)$$

From the difference scheme (5.3) it follows that

$$(E - \sigma\tau A_h)u_{j+1} = (E + \tau(1-\sigma)A_h)u_j,$$

or, briefly,

$$u_{j+1} = S u_j, \qquad u_0 = g.$$

Here, the transition operator is as follows:

$$S = E + (E - \sigma\tau A_h)^{-1}\tau A_h = E + \tau\overline{A}_h.$$

Hence,

$$u_N = S^N g = S^N P_h v_0 + S^N(g - P_h v_0)$$

and therefore (see (5.10), (5.7)) we have

$$\|\overline{A}_h^{1/2} u_N\| \le m + \|\overline{A}_h^{1/2}\| \|S\|^N \delta. \quad (5.20)$$

Since

$$\|S\|^N \le (1 + \tau\|\overline{A}_h\|)^N = (1 + \tau\|\overline{A}_h\|)^{T\|\overline{A}_h\|/\tau\|A_h\|} \le \exp(\|\overline{A}_h\|T), \quad (5.21)$$

estimates (5.11), (5.19)–(5.21) finally yield

$$\|\overline{A}_h^{1/2} w_N\| \le 2m + \|\overline{A}_h^{1/2}\| \exp(\|\overline{A}_h\|T)\delta \le 2m + c_2. \quad (5.22)$$

We apply Theorem 4.2 to the difference problem (5.13), (5.14) assuming that $H = H_h$, $A = A_+ = A_h \ge 0$, $A_- = 0$, and $\delta = q$. Then

$$\|w\|_{l_2(1,N;H_h)} \le \varepsilon\|\overline{A}_h^{1/2} w_N\| + c(\varepsilon)[q^{-1}\|f\|_{l_2(0,N-1;H_h)} + \|g - P_h v_0\|]$$
$$\le \varepsilon(2m + c_2) + Mc(\varepsilon)(\tau + h^p + \delta)$$

with some constant M dependent on σ, T, c, c_1, m, and q. Setting

$$\omega_m(\delta) = \inf_{\varepsilon > 0}(\varepsilon m + Mc(\varepsilon)\delta),$$

from estimate (4.17) (Chapter 5, Exercise 1.12) we obtain
$$\omega_m(\delta) = O(1/\ln \delta)^{-1})^{1/2}.$$
Therefore,
$$\|w\|_{l_2(1,N;H_h)} \leq \omega_{2m+c_2}(\tau + h^p + \delta) \to 0$$
for $\tau, h, \delta \to 0$. The theorem is proved. $\square$

Example 5.1. The mixed problem for the heat conduction equation with reverse time is an elementary example of an application of Theorem 5.1:
$$\frac{\partial v}{\partial t} = -\frac{\partial^2 v}{\partial x^2}, \quad x \in [0,1], \quad t \in [0,T],$$
$$v(x,0) = v_0(x), \quad x \in [0,1],$$
$$v(0,t) = v(1,t) = 0.$$

For the Hilbert space H we take the space $L_2(0,1)$. We denote by A the operator $-\mathrm{d}^2/\mathrm{d}x^2$, assuming that
$$D(A) = \{v(x) \in C^4[0,1] \mid v(0) = v(1) = 0\}.$$

The corresponding difference operator in the space of mesh functions $l_2(1, M-1) = H_h$ with norm
$$\|u\|_{H_h}^2 = h \sum_{m=1}^{M-1} |u^m|^2, \quad hM = 1,$$
is as follows:
$$(A_h u)^m = -\frac{u^{m+1} - 2u^m + u^{m-1}}{h^2}, \quad m = 1, \ldots, M-1,$$
$$u_0 = u_M = 0.$$

The operator of projection on the mesh, $P_h : u(x) \to u(mh) = u^m$, is correctly defined on the linear manifold $V = L_2(0,1) \cap C[0,1]$:
$$P_h(u) = u(mh) = u^m, \quad m = 1, \ldots, M-1.$$

It is easy to see that for any $v \in V$ condition (5.5) holds. A scalar product in H_h is defined as follows
$$\langle u, v \rangle = h \sum_{m=1}^{M-1} u^m v^m.$$

Chapter 6. Theory of stability of difference schemes

It is well known that

$$A_h^* = A_h \geq 0, \qquad \|A_h\| = \frac{4}{h^2}\cos^2(\pi h/2).$$

Since $v \in C^4[0,1]$, from the Taylor formula it follows that

$$(A_h P_h - P_h A)v = O(h^2), \tag{5.23}$$

i.e., the order of approximation $p = 2$.

For a fixed weight σ, the concordance condition (5.11) requires that the spacing h be much larger in order of magnitude than the error δ of the initial data. Indeed consider for simplicity the explicit scheme, $\sigma = 0$. We have $\overline{A}_h = A_h$ and estimate (5.11) takes the form

$$\|A_h^{1/2}\|\exp(T\|A_h\|)\delta \leq c_2. \tag{5.24}$$

In particular, by the inequality $\cos^2(\pi h/2) \geq 1/4$, from condition (5.24) we have

$$\exp(Th^{-2})\delta \leq c_2 \qquad \text{or} \qquad h \geq (T/\ln(c_2/\delta))^{1/2}$$

for the operator A_h from Example 5.1, where $h \leq 2/3$. The natural purpose to approximate the operator A by the operator A_h with accuracy equal to the error δ cannot be achieved because in this case we have

$$h = O(\delta^{1/2}) \tag{5.25}$$

according to (5.23).

We show that (5.25) will hold, if we assume the weight parameter $\sigma \in \mathbb{R}$ to be variable and determine it, for example, from the condition

$$F(\sigma) \equiv \|\overline{A}_h^{1/2} u_N(\sigma)\| = m + 1. \tag{5.26}$$

Then, under the assumption that σ obtained from (5.26) lies in the stability zone $\tau\sigma\|A_h\| \leq 1 - q$, from estimates (5.18), (5.19) and Theorem 4.2, we have

$$\|w\|_{l_2(1,N;H_h)} \leq \varepsilon(2m+1) + Mc(\varepsilon)[|\sigma|\tau + \tau + h^p + \delta].$$

Here the constant M depends only on T, c_1, m, and q and is independent of σ. Therefore, as before, we shall have

$$\|w\|_{l_2(1,N;H_h)} \leq \omega_{2m+1}(|\sigma|\tau + \tau + h^p + \delta) \to 0$$

for $h, \tau, \delta \to 0$ if we prove that

$$|\sigma|\tau \to 0. \tag{5.27}$$

Condition (5.27) is not evident since, generally speaking, the root σ of equation (5.26) tends to infinity. First, we show that $F(\sigma)$ defined in (5.26) monotonically increases. We have

$$F^2(\sigma) = \langle \overline{A}_h S^{2N} g, g \rangle.$$

By definition,

$$\overline{A}_h = (E - \sigma\tau A_h)^{-1} A_h, \qquad S = E + \tau \overline{A}_h,$$

and therefore, taking into account the formula

$$d\overline{A}_h/d\sigma = (E - \sigma\tau A_h)^{-2} \tau A_h^2 = \tau(\overline{A}_h)^2$$

we have

$$\frac{d}{d\sigma} F^2(\sigma) = \langle \tau(\overline{A}_h)^2 S^{2N} g, g \rangle + \langle \overline{A}_h 2N S^{2N-1} \tau^2 (\overline{A}_h)^2 g, g \rangle$$
$$= \langle \tau(\overline{A}_h)^2 S^{2N-1}[E + \tau\overline{A}_h + 2N\tau\overline{A}_h] g, g \rangle \geq \tau \langle \overline{A}_h g, \overline{A}_h g \rangle > 0.$$

This formula holds for $g \neq 0$ if $A_h > 0$, because $(E - \sigma\tau A_h) > 0$ for $\sigma\tau \|A_h\| \leq 1 - q$, $q \in (0,1)$. If we assume that

$$F(0) = \|A_h^{1/2}(E + \tau A_h)^N g\| > m + 1,$$

then, the root of equation (5.26) lies in the interval $(-\infty, 0)$ because $F(\sigma)$ monotonically increases. Estimate this root from below. We have

$$F(\sigma) = \|\overline{A}_h^{1/2} S^N g\| \leq \|\overline{A}_h^{1/2} S^N (g - P_h v_0)\| + \|\overline{A}_h^{1/2} S^N P_h v_0\|$$
$$\leq \|\overline{A}_h^{1/2}\| \|S\|^N \delta + m.$$

Estimate the norm of the operator $\overline{A}_h$, observing that $A_h = A_h^* \geq 0$. We have

$$\|\overline{A}_h\| = \sup_{\lambda \in \text{Sp } A_h} (\lambda/(1 - \sigma\tau\lambda));$$

and since

$$\frac{d}{d\lambda}\left(\frac{\lambda}{1 - \sigma\tau\lambda}\right) = \frac{1}{(1 - \sigma\tau\lambda)^2} > 0,$$

Chapter 6. Theory of stability of difference schemes

the maximum is attained for

$$\lambda = \|A_h\| \in \mathrm{Sp}\, A_h, \qquad \|\overline{A}_h\| = \frac{\|A_h\|}{1 - \sigma\tau\|A_h\|}.$$

In particular, for $\sigma \leq 0$ we have

$$\|\overline{A}_h\| \leq \|A_h\|, \qquad \|S\| \leq 1 + \frac{\tau\|A_h\|}{1 - \sigma\tau\|A_h\|},$$

$$\|S\|^N \leq (1 + \tau\|\overline{A}_h\|)^{T\|\overline{A}_h\|/\tau\|\overline{A}_h\|} \leq e^{T\|\overline{A}_h\|},$$

because $(1 + x)^{1/x} \leq e$. Hence, we obtain (see (5.28))

$$F(\sigma) \leq m + \exp\left\{\frac{1}{2}\ln\|A_h\| + \frac{T\|A_h\|}{1 - \sigma\tau\|A_h\|} - \ln\frac{1}{\delta}\right\}.$$

Now we find σ^* from the condition

$$\frac{1}{2}\ln\|A_h\| + \frac{T\|A_h\|}{1 - \sigma\tau\|A_h\|} = \ln\frac{1}{\delta}$$

i.e.,

$$\sigma^* = \frac{1}{\tau\|A_h\|} - \frac{T}{\tau\ln(1/(\delta\|A_h\|^{1/2}))}. \tag{5.28}$$

For $\sigma \leq \sigma^*$ we have $F(\sigma) \leq F(\sigma^*) \leq m + 1$. By assumption, $F(0) > m + 1$ and therefore $\sigma^* < 0$. In particular, the unique root of the equation $F(\sigma) = m + 1$ lies in the interval $[\sigma^*, 0)$. From equation (5.29) it follows that, generally speaking, $\sigma^* \to -\infty$, but $\sigma^*\tau \to 0$ only if $\|A_h\| \to \infty$ as $h \to 0$ and the parameters h and δ are coordinated so that

$$\sigma\|A_h\|^{1/2} \to 0, \qquad \delta, h \to 0.$$

Under these conditions, we have

$$|\sigma|\tau \leq |\sigma^*|\tau \to 0,$$

and the convergence of the difference scheme with weight σ determined from the condition $F(\sigma) = m + 1$ is proved. Thus, we have established the following theorem.

Theorem 5.2. *Suppose that the following conditions hold for the operator $A_h = A_h^* > 0$:*

$$\|(A_h P_h - P_h A)v\| \leq c_1 h^p, \qquad \|g - P_h v_0\| \leq \delta, \qquad \|A_h P_h v_t\| \leq c_1,$$

$$\|P_h v''\| \le c_1, \qquad \|\overline{A}_h^{1/2} P_h v_N\| \le m, \qquad \|\overline{A}_h^{1/2} S^N P_h v_0\| \le m,$$

$$\|A_h\| \to \infty, \qquad \delta\|A_h\|^{1/2} \to 0. \tag{5.29}$$

We set $\sigma = 0$ if $F(0) = \|A_h^{1/2} u_N(0)\| \le m+1$ and $\sigma = \sigma_0$ if $F(0) < m+1$, where σ_0 is the unique root of the equation

$$F(\sigma) = \|\overline{A}_h^{1/2} u_N(\sigma)\| = m+1.$$

Then the solution of the corresponding difference scheme with weights converges to the solution of the exact problem and

$$\|u - P_h v\|_{l_2(1,N;H_h)} \le \omega_{2m+1}(|\sigma_0|\tau + \tau + h^p + \delta),$$

where $\omega_m(\sigma) = O(1/\ln\delta^{-1})^{1/2}$.

Note that in Example 5.1 condition (5.30) holds for $h = O(\delta^{1/2})$ because $\delta\|A_h\|^{1/2} = O(\delta^{1/2})$. Since the function $F^2(\sigma)$ is convex downwards for $A_h > 0$, it is convenient to apply the Newton method for numerical determination of the root of the equation $F(\sigma) = m+1$ as follows. We apply this method for the equation $F^2(\sigma) - (m+1)^2 = 0$ taking $\sigma = 0$ as the initial approximation. Theorem 5.2 shows that $\sigma \to -\infty$ plays the role of the regularization parameter. If we need to find the smoothest solution $u(\sigma)$ within the given accuracy δ, we can determine the parameter $\sigma = \sigma_0$ from the condition

$$F(\sigma_0) = \min F(\sigma),$$

$$\sigma \in \Omega = \{\sigma \in \mathbb{R} \mid \sigma \ge 0, \quad \|g - \tilde{u}_N(\sigma)\| \le \delta\}, \tag{5.30}$$

where $\tilde{u}(\sigma)$ is a solution of the difference problem

$$\frac{\tilde{u}_{j+1} - \tilde{u}_j}{\tau} + A_h(\gamma \tilde{u}_{j+1} + (1-\gamma)\tilde{u}_j) = 0, \tag{5.31}$$

$$\tilde{u}_0 = u_N(\sigma) \tag{5.32}$$

which is inverse to the problem (5.3), (5.4).

Here $u(\sigma)$ is the solution of the difference problem (5.3), (5.4) and the parameter γ is determined from the stability condition for the difference problem (5.32), (5.33). By Theorem 3.3, we may take, for example, γ such that $E - \tau(1-2\gamma)A_h \ge 0$, in particular, $\gamma = 1/2$.

In this algorithm, as in the algorithm from Theorem 5.2, a kind of automatic adjustment of the difference scheme takes place. The idea of the

Chapter 6. Theory of stability of difference schemes

algorithm (5.31)–(5.33) is close to the algorithm of choosing the parameter α by the residual (see Section 3, Chapter 5). However, this algorithm is distinct from the algorithm of Chapter 5 at least by the fact that different finite-difference schemes are used for the solution of the direct and inverse problems. Theorems 3.2 and 3.3 show that, generally speaking, we cannot use the same difference scheme with weight for solving both direct and inverse problems because difference schemes with $\sigma > 0$ are preferable for well-posed problems and schemes with $\sigma \leq 0$ are preferable for ill-posed problems.

We give heuristic explanation of the regularizing influence of the parameter σ. We consider the difference scheme with weight

$$\frac{u(t+\tau) - u(\tau)}{\tau} - A(\sigma u(t+\tau) + (1-\sigma)u(t)) = 0,$$

assuming for simplicity that time t is continuous and $A > 0$. Since formally we have

$$\hat{u}(t) = u(t+\tau) = \exp(\tau D)u(t),$$

where $D = \partial/\partial t$, we see that

$$\hat{u} = u + \tau \frac{\partial u}{\partial t} + O(\tau^2),$$

$$u_t = \frac{\partial u}{\partial t} + \frac{\tau}{2}\frac{\partial^2 u}{\partial t^2} + O(\tau^2),$$

$$u_t - A(\sigma\hat{u} + (1-\sigma)u) = \frac{\partial u}{\partial t} + \frac{\tau}{2}\frac{\partial^2 u}{\partial t^2} - Au - \sigma\tau A\frac{\partial u}{\partial t} + O(\tau^2).$$

If u is a solution of the equation

$$\frac{\partial u}{\partial t} = Au, \tag{5.33}$$

then $\partial^2 u/\partial t^2 = A^2 u$ and therefore

$$u_t - A(\sigma\hat{u} + (1-\sigma)u) = \frac{\partial u}{\partial t} - Au - \tau(\sigma - 1/2)A^2 u + O(\tau^2).$$

Hence, the difference scheme with weight approximates the problem (5.34) with order $O(\tau(\sigma - 1/2))$ and the problem

$$\frac{\partial u}{\partial t} - Au - \tau(\sigma - 1/2)A^2 u = 0 \tag{5.34}$$

with order $O(\tau^2)$.

For $\sigma \geq 1/2$, the problem (5.35) is more unstable than (5.34) and therefore it makes no sense to use the schemes with $\sigma > 1/2$. In contrast, for $\sigma < 1/2$, the problem (5.35) is well-posed because it has viscous term $\tau(\sigma - 1/2)A^2 u$.

Thus, if we use the schemes with weights, the appearing approximation viscosity regularizes the initial ill-posed problem. In this case, evidently, it is not necessary to introduce viscosity in the initial differential equation

$$\frac{\partial u}{\partial t} - Au + \alpha A^2 u = 0, \qquad \alpha > 0,$$

as it is assumed in the known method of quasi-inversion (see Lattès and Lions, 1967). Introducing artificial viscosity is undesirable also because in this case it is necessary to assume additional smoothness of the differential problem and realization of additional boundary conditions.

Smoothing influence of the parameter $\sigma \to -\infty$ is evident also from the formula for the transition operator S

$$S = E + \tau A_h (E - \sigma \tau A_h)^{-1}.$$

Exercise 5.1. Give rigorous substantiation for the algorithm of choosing σ on the basis of conditions (5.31)–(5.33) and prove the convergence theorem.

Exercise 5.2. Use the estimate of l-stability in the interior part of the mesh Z_0^N obtained in Theorem 1.1 to prove the corresponding convergence theorems and compare the rate of convergence with the estimates of Theorems 5.1, 5.2.

Exercise 5.3. Generalize Theorems 5.1, 5.2 to the case of arbitrary constant operator $A_h = A_h^*$.

Exercise 5.4. Write the Euler equation for the problem (5.3)–(5.4) (see (3.17), Chapter 5) and compare the variational method and the algorithms presented in this chapter in the number of arithmetic operations.

6.6. FINITE STABILITY OF TWO-LAYER SCHEMES OF THE CANONICAL FORM

In this section, we generalize Theorem 3.5 to the case of two-layer canonical schemes

$$Pu = Bu_t + Au = f$$

Chapter 6. Theory of stability of difference schemes

without the assumption of self-adjointness of the operators A and B. So, we suppose that $A, B \in \mathcal{L}(H)$ are independent of j and, moreover,

$$A = (A + A^*)/2 + i(A - A^*)/2i = \operatorname{Re} A + i \operatorname{Im} A,$$

$$B = (B + B^*)/2 + i(B - B^*)/2i = \operatorname{Re} B + i \operatorname{Im} B,$$

Making the change of variables similar to that of Section 3:

$$u = \frac{1}{\psi}v, \qquad u_t = \frac{1}{\psi}(v_t - s\varphi_t \hat{v}),$$

we get

$$I = \|Pu\|_s^2 = \tau \sum_{j=0}^{N-1} \|B(v_t - s\varphi_t \hat{v}) + Av\|^2$$

$$= \tau \sum_{j=0}^{N-1} \{\|Bv_t\|^2 + \|Av - s\varphi_t B\hat{v}\|^2 + 2\operatorname{Re}\langle Bv_t, Av\rangle$$

$$- s\varphi_t \, 2\operatorname{Re}\langle Bv_t, B\hat{v}\rangle\}$$

$$= I_1 + I_2 + I_3 + I_4. \qquad (6.1)$$

Since

$$\partial \langle Bv, Av\rangle = \langle Bv_t, Av\rangle + \langle B\hat{v}, Av_t\rangle$$
$$= 2\operatorname{Re}\langle Bv_t, Av\rangle - \langle Av, Bv_t\rangle + \langle Bv, Av_t\rangle + \tau \langle Bv_t, Av_t\rangle$$
$$= 2\operatorname{Re}\langle Bv_t, Av\rangle + \langle (A^*B - B^*A)v, v_t\rangle + \tau \langle v_t, B^*Av_t\rangle,$$

taking once more the real part of both sides of the equality, we obtain

$$2\operatorname{Re}\langle Bv_t, Av\rangle = \partial \langle \operatorname{Re}(A^*B)v, v\rangle + 2\operatorname{Re}\langle i\operatorname{Im}(B^*A)v, v_t\rangle - \tau \langle \operatorname{Re}(B^*A)v_t, v_t\rangle. \qquad (6.2)$$

Here we have used the fact that

$$\operatorname{Re}\langle Au, u\rangle = \langle \operatorname{Re} Au, u\rangle, \qquad \forall A \in \mathcal{L}(H).$$

Summing by parts in (6.2) and observing that $v_0 = v_N = 0$, we have

$$I_3 = \tau \sum_{j=0}^{N-1} \{2\operatorname{Re}\langle i\operatorname{Im}(B^*A)v, v_t\rangle - \tau \langle \operatorname{Re}(B^*A)v_t, v_t\rangle\}.$$

Further, since

$$\partial\langle Bv, Bv\rangle = \langle Bv_t, B\hat{v}\rangle + \langle Bv, Bv_t\rangle = 2\operatorname{Re}\langle Bv_t, B\hat{v}\rangle - \tau\langle Bv_t, Bv_t\rangle,$$

$$2\operatorname{Re}\langle Bv_t, B\hat{v}\rangle = \partial\langle Bv, Bv\rangle + \tau\langle Bv_t, Bv_t\rangle,$$

summing by parts once more, we have

$$I_4 = \tau \sum_{j=0}^{N-1} -s\varphi_t\{\partial\langle Bv, Bv\rangle + \tau\langle Bv_t, Bv_t\rangle\}$$

$$= \tau \sum_{j=0}^{N-1}\{s\varphi_{t\bar{t}}\langle B^*Bv, v\rangle - s\tau\varphi_t\langle B^*Bv_t, v_t\rangle\}.$$

Now, we transform I_2. By definition (6.1),

$$I_2 = \tau \sum_{j=0}^{N-1}\{\langle A^*Av, v\rangle + s^2\varphi_t^2\|B\hat{v}\|^2 - s\varphi_t \cdot 2\operatorname{Re}\langle Av, B\hat{v}\rangle\}. \tag{6.3}$$

For any numerical functions of the discrete argument $x = x_j$, $y = y_j$, $j = 0, \ldots, N$ with the condition $x_0 = x_N = 0$, we have the identity

$$\sum_{j=0}^{N-1}\hat{x}y = \sum_{j=0}^{N-1}x\check{y}.$$

In other words, the shift operator may be transferred from one multiplier to another, changing the shift direction. Therefore, since $\hat{\varphi}_t = \varphi_{\bar{t}}$, we get

$$\tau\sum_{j=0}^{N-1} s^2\varphi_t^2\|B\hat{v}\|^2 = \tau\sum_{j=0}^{N-1} s^2\varphi_t^2\|Bv\|^2. \tag{6.4}$$

Now we transform $2\operatorname{Re}\langle Av, B\hat{v}\rangle$:

$$2\operatorname{Re}\langle Av, B\hat{v}\rangle = 2\operatorname{Re}\langle Av, Bv\rangle + \tau\, 2\operatorname{Re}\langle Av, Bv_t\rangle$$
$$= \langle 2\operatorname{Re}(B^*A)v, v\rangle + \tau\, 2\operatorname{Re}\langle\operatorname{Re}(B^*A)v, v_t\rangle$$
$$+ \tau\, 2\operatorname{Re}\langle i\operatorname{Im}(B^*A)v, v_t\rangle. \tag{6.5}$$

In the second term, the operator $D = \operatorname{Re}(B^*A)$ is self-adjoint. Therefore, taking into account the formula

$$2\operatorname{Re}\langle Dv, v_t\rangle = \partial\langle Dv, v\rangle - \tau\langle Dv_t, v_t\rangle, \tag{6.6}$$

Chapter 6. Theory of stability of difference schemes

we may sum this term by parts. As a result, (6.3), (6.4)–(6.6) yield

$$I_2 = \tau \sum_{j=0}^{N-1} \{\langle A^*Av, v\rangle + s^2\varphi_{\bar{t}}^2\langle B^*Bv, v\rangle - s\varphi_t\langle 2\operatorname{Re}(B^*A)v, v\rangle$$
$$- s\tau\varphi_t\, 2\operatorname{Re}\langle i\operatorname{Im}(B^*A)v, v_t\rangle + s\tau\varphi_{t\bar{t}}\langle \operatorname{Re}(B^*A)v, v\rangle$$
$$+ s\tau^2\varphi_t\langle \operatorname{Re}(B^*A)v_t, v_t\rangle\}.$$

Substituting the expressions for I_2, I_3, and I_4 into formula (6.1), we obtain

$$\|Pu\|_s^2 = \tau\sum_{j=0}^{N-1}\{\langle \mathcal{A}v, v\rangle + 2\operatorname{Re}\langle \mathcal{B}v, v_t\rangle + \langle \mathcal{C}v_t, v_t\rangle\}. \tag{6.7}$$

Here the self-adjoint operators $\mathcal{A}$ and $\mathcal{C}$ and the skew-self-adjoint operator $\mathcal{B} = -\mathcal{B}^*$ are defined as follows

$$\mathcal{A} = A^*A + s^2\varphi_{\bar{t}}^2 B^*B - (s\varphi_t - s\tau/2\,\varphi_{t\bar{t}})2\operatorname{Re}(B^*A) + s\varphi_{t\bar{t}}B^*B,$$
$$\mathcal{B} = i\operatorname{Im}(B^*A)(1 - s\tau\varphi_t),$$
$$\mathcal{C} = s\tau^2\varphi_t\operatorname{Re}(B^*A) - \tau\operatorname{Re}(B^*A) - s\tau\varphi_t B^*B + B^*B. \tag{6.8}$$

We use the identity

$$(A - \alpha B)^*(A - \alpha B) = A^*A + \alpha^2 B^*B - 2\alpha\operatorname{Re}(B^*A),$$

which holds for any real α. Setting

$$\alpha = s\varphi_t - s\frac{\tau}{2}\cdot\varphi_{t\bar{t}} = \frac{s(\varphi_t + \varphi_{\bar{t}})}{2} = s\partial_0\varphi,$$

in this identity (the last equality is the definition of the operator ∂_0) we transform the expression for the operator $\mathcal{A}$ to the form

$$\mathcal{A} = (A - s\partial_0\varphi B)^*(A - s\partial_0\varphi B) + s\gamma B^*B, \tag{6.9}$$

where

$$\gamma = \varphi_{t\bar{t}} + s(\varphi_{\bar{t}}^2 - (\partial_0\varphi)^2). \tag{6.10}$$

In its turn,

$$\mathcal{C} = (1 - s\tau\varphi_t)\{B^*B - \tau\operatorname{Re}(B^*A)\}. \tag{6.11}$$

Thus, we have proved the following lemma.

Lemma 6.1. *Identity (6.7) with* $u = \psi^{-1}v$ *and the operators* $\mathcal{A}$, $\mathcal{B}$, *and* $\mathcal{C}$ *defined by formulas (6.8)–(6.11) holds for all* $u \in C_0(Z_0^N)$.

Lemma 6.1 implies the following lemmas.

Lemma 6.2. Suppose for all $s \geq s_0$ and some $\delta > 0$ the following inequality holds:
$$\begin{pmatrix} A - s\delta E, & -B \\ B, & C \end{pmatrix} \geq 0.$$
Then
$$s\|u\|_s^2 \leq c\|Pu\|_s^2, \qquad c = \delta^{-1} \qquad (6.12)$$
for all $u \in C_0(Z_0^N)$ and $s \geq s_0$.

Lemma 6.3. Suppose that for the operators A, B, and C the following inequalities hold for $s \geq s_0$:
$$C > 0, \quad A - B^*C^{-1}B \geq s\delta E, \qquad \delta > 0.$$
Then estimate (6.12) holds.

The proof of Lemmas 6.2, 6.3 follows from Lemma 6.1 and the identities
$$\langle Av, v \rangle + 2\operatorname{Re}\langle Bv, v_t \rangle + \langle Cv_t, v_t \rangle = \begin{pmatrix} v \\ v_t \end{pmatrix}^T \begin{pmatrix} A, & -B \\ B, & C \end{pmatrix} \begin{pmatrix} v \\ v_t \end{pmatrix}$$
$$= \langle C^{-1}(Cv_t + Bv), Cv_t + Bv \rangle + \langle (A - B^*C^{-1}B)v, v \rangle.$$

Lemmas 6.1–6.3 yield a series of sufficient conditions of stability in terms of the initial operators A and B (see Section 3). We restrict ourselves to the following two theorems.

Theorem 6.1. Suppose $B^{-1} \in \mathcal{L}(H)$ and for some $s_0 > 0$ the following inequalities hold:
$$B^*B - \tau\operatorname{Re}(A^*B) \geq \frac{1}{s_0}\operatorname{Im}(A^*B)(B^*B)^{-1}\operatorname{Im}(A^*B), \qquad (6.13)$$
$$\varphi_{t\bar{t}} + s_0\tau\varphi_t \geq 2, \qquad -\varphi_t > 0. \qquad (6.14)$$
Then
$$s\|Bu\|_s^2 \leq \|Bu_t + Au\|_s^2$$
for all $s \geq s_0$ and $u \in C_0(Z_0^N)$.

Chapter 6. Theory of stability of difference schemes

Proof. Since $\varphi_{t\bar{t}} > 0$ and $-\varphi_t > 0$, we obtain

$$\varphi_{\bar{t}}^2 - (\partial_0\varphi)^2 = -\tau\varphi_{t\bar{t}}(\varphi_t + 3\varphi_{\bar{t}})/4 > 0;$$

and therefore the function γ from (6.10) is greater than $\varphi_{t\bar{t}}$. Hence, from the definition of the operator $\mathcal{A}$ (see (6.9)) it follows that

$$\mathcal{A} \geq (A - s\partial_0\varphi B)^*(A - s\partial_0\varphi B) + s\varphi_{t\bar{t}} B^*B. \tag{6.15}$$

Further,

$$2\operatorname{Re}\langle \mathcal{B}v, v_t\rangle = 2\operatorname{Re}\langle \mathcal{B}B^{-1}Bv, v_t\rangle = 2\operatorname{Re}\langle Bv, (\mathcal{B}B^{-1})^*v_t\rangle$$
$$\leq \varepsilon\|Bv\|^2 + \|(\mathcal{B}B^{-1})^*v_t\|^2/\varepsilon,$$

which implies

$$\langle \mathcal{A}v, v\rangle + 2\operatorname{Re}\langle \mathcal{B}v, v_t\rangle + \langle \mathcal{C}v_t, v_t\rangle$$
$$\geq \langle (\mathcal{A} - \varepsilon B^*B)v, v\rangle + \langle \{\mathcal{C} - \varepsilon^{-1}(\mathcal{B}B^{-1})(\mathcal{B}B^{-1})^*\}v_t, v_t\rangle.$$

Setting $\varepsilon = (1 - s\tau\varphi_t)s_0$ and taking into account (6.8), (6.11), (6.13)–(6.15), we obtain

$$\mathcal{A} - \varepsilon B^*B \geq (A - s\partial_0\varphi B)^*(A - s\partial_0\varphi B) + s(\varphi_{t\bar{t}} - s_0/s + s_0\tau\varphi_t)B^*B$$
$$\geq (A - s\partial_0\varphi B)^*(A - s\partial_0\varphi B) + sB^*B, \tag{6.16}$$

$$\mathcal{C} - \varepsilon^{-1}\mathcal{B}B^{-1}(B^{-1})^*\mathcal{B}$$
$$= (1 - s\tau\varphi_t)\{B^*B - \tau\operatorname{Re}(B^*A) - (\operatorname{Im}(B^*A))(B^*B)^{-1}\operatorname{Im}(B^*A)/s_0\} \geq 0.$$

These estimates and Lemma 6.1 yield

$$s\|Bu\|_s^2 + \|(A - s\partial_0\varphi B)u\|_s^2 \leq \|Bu_t + Au\|_s^2. \tag{6.17}$$

Here we have used the fact that

$$\|u\|_s^2 = \|v\|_0^2 = \tau\sum_{j=0}^{N-1}\|v_j\|^2, \qquad v = \psi u.$$

Neglecting the second term in estimate (6.17), we obtain the assertion of the theorem. □

The term $\|(A-s\partial_0\varphi B)u\|_s^2$ can be used to estimate $\|Au\|_s$ and $\|Bu_t\|_s$ via the right-hand side of inequality (6.17). Indeed, for any number $c>0$ we have

$$\frac{c}{s}\|Au\|_s^2 = \frac{c}{s}\|Au - s\partial_0\varphi Bu + s\partial_0\varphi Bu\|_s^2$$
$$\leq \frac{c}{s}\{2\|Au - s\partial_0\varphi Bu\|_s^2 + 2s^2(\partial_0\varphi)^2\|Bu\|_s^2\},$$

which implies

$$\frac{2c}{s}\|(A-s\partial_0\varphi B)u\|_s^2 \geq \frac{c}{s}\|Au\|_s^2 - 2cs(\partial_0\varphi)^2\|Bu\|_s^2.$$

Suppose

$$\varphi_{t\bar{t}} - \frac{s_0}{s} + s_0\tau\varphi_t - 2c(\partial_0\varphi)^2 \geq 1$$

for $s \geq s_0$, or, briefly, $\varphi_{t\bar{t}} + s_0\tau\varphi_t - 2c(\partial_0\varphi)^2 \geq 2$ and the number s_0 is so large that $2c/s_0 \leq 1$. Then we have $2c/s \leq 1$ for $s \geq s_0$ and, acting as in the proofs of estimates (6.16) and (6.17), we have

$$s\|Bu\|_s^2 + \frac{c}{s}\|Au\|_s^2 \leq \|Bu_t + Au\|_s^2.$$

Hence,

$$\frac{c}{s}\|Bu_t\|_s^2 = \frac{c}{s}\|Bu_t + Au - Au\|_s^2$$
$$\leq \frac{2c}{s}\|Bu_t + Au\|_s^2 + \frac{2c}{s}\|Au\|_s^2 \leq 2\|Bu_t + Au\|_s^2.$$

Finally,

$$s\|Bu\|_s^2 + \frac{c}{s}(\|Au\|_s^2 + \|Bu_t\|_s^2) \leq 3\|Bu_t + Au\|_s^2. \qquad (6.18)$$

Thus, we have proved the following theorem.

Theorem 6.2. *Suppose* $B^{-1} \in \mathcal{L}(H)$,

$$\varphi_{t\bar{t}} + s_0\tau\varphi_t - 2c(\partial_0\varphi)^2 \geq 2, \qquad 2c < s_0$$

and inequality (6.13) holds. Then estimate (6.18) holds for all $s \geq s_0$ and $u \in C_0(Z_0^N)$.

Chapter 6. Theory of stability of difference schemes

6.7. CONDITIONS OF STABILITY IN TERMS OF THE TRANSITION OPERATOR

A two-layer scheme in terms of the transition operator is as follows:

$$(Pu)_j = \frac{u_{j+1} - R_j u_j}{\tau} = f_j,$$

or

$$u_{j+1} = R_j u_j + \tau f_j. \tag{7.1}$$

Here $R_j \in \mathcal{L}(H)$ is the operator of transition from the jth layer to the $(j+1)$th layer. The classical stability condition in terms of the transition operator is as follows:

$$\|R_j\| \leq 1 + c_0 \tau, \qquad j = 0, \ldots, N-1, \tag{7.2}$$

where the number c_0 is independent of τ. It is known that condition (7.2) implies the estimate of stability with respect to the right-hand side and the initial data. For $u_0 = 0$ the latter estimate has the form

$$\|u_n\| \leq M\tau \sum_{j=0}^{n-1} \|(Pu)_j\|, \qquad n = 1, \ldots, N. \tag{7.3}$$

The question arises of whether the scheme (7.1) is finitely stable under condition (7.2). In other words, does our definition of finite stability extend the classical definition of the stability condition? The following theorem gives the positive answer to this question.

Theorem 7.1. *Let estimate* (7.3) *with* $u_0 = 0$ *hold for the difference scheme* P. *Then for all* $s > 0$, $u : Z_0^N \to H$, *the following estimate holds*

$$s\|u\|_s \leq c\|Pu\|_s, \tag{7.4}$$

where $c = M/\mu$, $\mu = \min_j(-\varphi_t) > 0$. *In particular, the scheme is finitely stable.*

Proof. For $u : Z_0^N \to H$, $u_0 = 0$, from the definition of $\|u\|_s$ and estimate (7.3) we have $(f = Pu)$

$$\|u\|_s^2 \leq \tau \sum_{n=1}^{N-1} \Psi_n^2 \left\{ M\tau \sum_{j=0}^{n-1} \|f_j\| \right\}^2 = \tau \sum_{n=1}^{N-1} \left(\sum_{j=0}^{n-1} (\Psi_n/\Psi_j) \Psi_j \|f_j\| \right)^2 M^2 \tau^2$$

$$\leq \tau \sum_{n=1}^{N-1} \left\{ \sum_{j=0}^{n-1} (1 + \mu\tau s)^{-(n-j)} \Psi_j \|f_j\| \right\}^2 M^2 \tau^2 \tag{7.5}$$

because in view of estimate (1.10) $\Psi_n/\Psi_j \leq k_{n-j}$, where $k_j = (1+\mu\tau s)^{-j}$.

Applying the Cauchy-Schwartz-Bunyakovskii inequality to the interior sum of (7.5), we obtain

$$\left\{\sum_{j=0}^{n-1} k_{n-j}^{1/2} k_{n-j}^{1/2} \Psi_j \|f_j\|\right\}^2 \leq K \sum_{j=0}^{n-1} k_{n-j} \Psi_j^2 \|f_j\|^2, \qquad (7.6)$$

where $K = \sum_{j=0}^{N-1} k_j$. Changing the order of summation in (7.5) and using (7.6), we obtain the estimate

$$\|u\|_s^2 \leq (KM\tau)^2 \|f\|_s^2.$$

By definition, we have

$$K = \sum_{j=1}^{N-1} q^j \leq \frac{q}{1-q} = \frac{1}{s\tau\mu},$$

since $q = (1+\mu\tau s)^{-1}$. Hence, $(KM\tau)^2 = (M/\mu)^2 s^{-2}$ and therefore

$$s^2 \|u\|_s^2 \leq c^2 \|Pu\|^2.$$

Extracting the square root, we obtain (7.4). The theorem is proved. □

Since condition (7.2) implies estimate (7.3) with the constant $M = \exp(c_0 T)$, Theorem 7.1 yields the following assertion.

Corollary 7.1. The difference scheme (7.1) is finitely stable under condition (7.2). More precisely, for all $u : Z_0^N \to H$, $u_0 = 0$, estimate (7.4) holds.

Note that we have not used the Hilbert structure of the space H in the proof of Theorem 7.1. Therefore, H can be assumed to be an arbitrary normed space in this theorem.

Now, we shall find out the extent to which the notion of finite stability is more general than the classical definition (7.2) or, more precisely, how we can make conditions (7.2) weaker. First, suppose that the transition operator R is independent of j. Setting $R = E - \tau A$, we write the difference scheme (7.1) in the canonical form $u_t + Au = f$ with $B = E$. As far as $\operatorname{Re} R = E - \tau \operatorname{Re} A$, $\operatorname{Im} R = -\tau \operatorname{Im} A$, the finite stability condition from Theorem 6.1 can be written in the form of the inequality

$$\tau^2 \operatorname{Re} R \geq (\operatorname{Im} R)^2 / s_0, \qquad (7.7)$$

where s_0 is an arbitrary number that can be as large as desired. In particular, if $R = R^*$, then condition (7.7) is transformed into the condition $R \geq 0$. For the normal operator R, condition (7.2) implies that the spectrum of the operator R lies in the disk $|\lambda| \leq 1 + c_0\tau$. Condition (7.7) adds to this set a tail in the form of the interior of the parabola in the complex plane:

$$\tau^2 \operatorname{Re} \lambda \geq (\operatorname{Im} \lambda)^2 / s_0.$$

For the variable transition operator R, applying Theorem 3.10 with $\sigma = 0$, $A = (\operatorname{Re} R - E)/\tau$, $B = (\operatorname{Im} R)/\tau$, we obtain the following sufficient conditions of finite stability

$$\operatorname{Re} R \geq E, \qquad \|\operatorname{Im} R\| \leq c\tau,$$

$$-\partial \operatorname{Re} R \leq c(\operatorname{Re} R - E),$$

$$-i[\operatorname{Re} R, \operatorname{Im} R] \leq \tau c_1 (\operatorname{Re} R - E).$$

Here, as usual, $\partial u = u_t$.

Investigation of the stability of three-layered schemes is reduced to investigation of the system of two two-layer schemes. For example, we may consider the following three-layered scheme

$$u_{tt} - Au = f.$$

More precisely,

$$\frac{u_{j+2} - 2u_{j+1} + u_j}{\tau^2} - Au_j = f_j, \qquad (7.8)$$

where A is a constant self-adjoint operator.

We set

$$u_{j+1} - Ru_j = \tau v_j, \qquad (7.9)$$

$$v_{j+1} - Sv_j = \tau f_j \qquad (7.10)$$

and choose the operators R and S so that the system (7.9), (7.10) is transformed into equation (7.8) after excluding v. Excluding from v_j from (7.10) by means of (7.9), we obtain

$$u_{j+2} - (R+S)u_{j+1} + Sru_j = \tau^2 f_j,$$

which implies

$$R + S = 2E, \qquad (7.11)$$

$$SR = E - \tau^2 A. \qquad (7.12)$$

We represent the operator A as the difference of two nonnegative commuting operators $A_\pm$

$$A = A_+ - A_-, \quad (A_\pm)^* = (A_\pm) \geq 0, \quad A_+ = QA, \quad A_- = (Q-E)A,$$

where Q is the orthogonal projector which projects the space onto the subspace of eigenfunctions corresponding to the positive part of the spectrum of A. Since $A_\pm \geq 0$, the nonnegative operators $\sqrt{A_\pm}$ are determined uniquely. As a definition, we set

$$\sqrt{A} = \sqrt{A_+} + i\sqrt{A_-}.$$

By the construction of $A_\pm$ we have

$$A_+ A_- = A_- A_+ = 0$$

and therefore

$$E - \tau^2 A = (E - \tau\sqrt{A})(E + \tau\sqrt{A}).$$

Now, setting $S = E - \tau\sqrt{A}$ and $R = E + \tau\sqrt{A}$, we find the solution of the system of equations (7.11), (7.12). Suppose that the operators R and S constructed above satisfy condition (7.7). In terms of the operators $A_\pm$ it means that

$$E + \tau\sqrt{A_+} \geq A_-/s_0, \quad E - \tau\sqrt{A_+} \geq A_-/s_0.$$

As far as difference schemes (7.9) and (7.10) are finitely stable, by Theorem 7.1 we have

$$s\|u\|_s^2 \leq \|v\|_s^2 \leq s^{-1}\|f\|_s^2.$$

As a result, we obtain

$$s^2 \|u\|_s^2 \leq \|f\|_s^2,$$

if $u_0 = u_N = 0$ and $v_0 = v_N = 0$. Or, briefly, $u_0 = u_1 = 0$, $u_{N-1} = u_N = 0$. In this case, we assume that $s \geq s_0$ and the weight function φ that appears in the definition of the norm $\|\cdot\|_s$ satisfies the assumptions of Theorem 6.1. In the general case the three-layered scheme of the form

$$Pu = u_{tt} + Bu_t + Au = f$$

is called *finitely stable* if there exists a number M independent of s, τ, $\|A\|$, and $\|B\|$, such that for all functions

$$u \in C_0^2(Z_0^N) = \{u : Z_0^N \to H \mid u_0 = u_1 = u_{N-1} = u_N = 0\}$$

Chapter 6. Theory of stability of difference schemes

and for any $s \geq s_0$ the following estimate holds:

$$s^2 \|u\|_s^2 \leq M \|Pu\|_s^2$$

Using the above-mentioned method of reducing a three-layered scheme to the system of two-layer schemes, it is easy to find the sufficient conditions for its finite stability and formulate the corresponding convergence theorems.

Exercise 7.1. Suppose $A, B \in \mathcal{L}(H)$, $A = A^* > 0$, $B > 0$, $B - \tau A/2 \geq 0$, $-\varphi_t \geq 1$. Prove that

$$s\|u\|_{A,s} \leq \|B^{-1}(Bu_t + Au)\|_{A,s},$$

for all $s > 0$, where $u : Z_0^N \to H$, $u_0 = 0$, and

$$\|u\|_{A,s}^2 = \tau \sum_{j=0}^{N-1} \Psi^2 \|u_j\|_A^2, \qquad \|u\|_A^2 = \langle Au, u \rangle.$$

Exercise 7.2. Reducing the Cauchy problem for the Laplace equation $u_{xx} + u_{yy} = 0$ to the Cauchy problem for the Cauchy-Riemann system, investigate the stability of its difference scheme with weights.

Exercise 7.3. Consider the difference scheme with weights for the Cauchy problem

$$\frac{\partial u}{\partial t} = \frac{1}{i} \frac{\partial u}{\partial x} + f, \qquad i^2 = -1.$$

Set $u_m^n = u(mh, n\tau)$, $h, \tau > 0$. For the space H we take the space of mesh functions $u : \mathbb{Z} \to \mathbb{C}$,

$$\langle u, v \rangle = h \sum_{m=-\infty}^{\infty} u_m \bar{v}_m.$$

We define the operator A by the formula

$$(Au)_m = \frac{1}{i} \frac{u_{m+1} - u_{m-1}}{2h}.$$

Then the scheme with weights takes the form

$$\frac{u_m^{n+1} - u_m^n}{\tau} = \sigma (Au)_m^{n+1} + (1 - \sigma)(Au)_m^n + f_m^n.$$

Show that for finite stability it is necessary that

$$|\sigma|\frac{\tau}{h} < 1.$$

If, in addition,

$$1 + \frac{\tau}{h}(1 - 2\sigma)\sin h\lambda \geq 0, \qquad \forall \lambda \in [-\frac{\pi}{h}, \frac{\pi}{h}],$$

then the scheme with weights is finitely stable.

Investigate the finite stability of the following difference schemes

$$\frac{u_m^{n+1} - u_m^n}{\tau} = \frac{u_{m+1}^n - u_{m-1}^n}{i2h} - \tau\frac{u_{m+1}^n - 2u_m^n + u_{m-1}^n}{2h^2},$$

$$\frac{(u_{m+1}^{n+1} + u_m^{n+1}) - (u_{m+1}^n + u_m^n)}{2\tau} = \frac{(u_{m+1}^{n+1} + u_{m+1}^n) - (u_m^{n+1} + u_n^n)}{i2h}$$

and compare the obtained conditions with the stability conditions of Chudov (1967, p. 51–54).

6.8. REMARKS AND REFERENCES

The results of this chapter are obtained by the author (see Bukhgeim (1983a), (1986)). There exist other approaches to the study of these problems. As was noted in Introduction, the problem of developing the theory of stability of difference schemes for ill-posed boundary value problems was first stated by Chudov (1962, 1967). He used the method of Fourier transform for this purpose. The approach based on the definition of ρ-stability introduced by Samarskii (1977) is developed by Vabishchevich (1986, Chapter 3). Some methods are presented in Marchuk (1980). The quasi-inversion method was investigated in details in Lattès and Lions (1967).

Further references can be found in Lattès and Lions (1967), Marchuk (1980), Bukhgeim (1986a), and Vabishchevich, et al. (1986).

Appendix A

In Chapter 3, we saw that investigation of the Cauchy problem for the Schrödinger equation with operator coefficients is very useful when proving the uniqueness of solutions of multidimensional inverse problems. The purpose of this appendix is to prove that investigation of the operator variant of boundary value problems for the Cauchy–Riemann system or the Laplace equation and their generalizations is closely related to the problems of computational tomography. This is the problem of reconstruction of the unknown function by its integrals along a certain set of curves or surfaces. Such connection allows us to obtain new variants of inversion formulas which take into account the geometric characteristics of the domain Ω where the sought function is defined. In this case, we reduce the numerical solution of the problems mentioned above to the solution of elliptic boundary value problems. This makes it possible to use advanced methods of computational mathematics. For simplicity we shall illustrate this approach by the example of the plane Radon problem in a bounded domain Ω with smooth boundary. It is required to find a continuous function $a(x, y)$ defined in the domain Ω by the integrals of this function along various straight lines that intersect this domain. In the differential form, this problem is reduced to finding the right-hand side of the equation

$$Pu \equiv \langle \nu, \nabla u \rangle = a(x, y) \qquad (A.1)$$

by the values of the solution $u(x, y, \alpha)$ on the boundary

$$u|_\Sigma = f, \qquad \Sigma = \partial \Omega \times [-\pi, \pi]. \qquad (A.2)$$

Here $\nu = (\cos \alpha, \sin \alpha)$ is a unit vector, ∇ is the gradient with respect to the variables x and y, $\langle \cdot, \cdot \rangle$ is a scalar product in $\mathbb{R}^2$, u is a real smooth function which is 2π-periodic with respect to α. Expand the function u in the Fourier series with respect to α:

$$u(x, y, \alpha) = \sum_{n=-\infty}^{\infty} u_n(x, y) e^{in\alpha}. \qquad (A.3)$$

Since u is a real function, the Fourier coefficients u_n are complex-conjugate

$$u_n = \overline{u}_{-n}. \qquad (A.4)$$

Since
$$\cos\alpha = \frac{e^{i\alpha} + e^{-i\alpha}}{2}, \qquad \sin\alpha = \frac{e^{i\alpha} - e^{-i\alpha}}{2i},$$

substituting formula (3) into equation (1) and multiplying by $2e^{-i\alpha}$, we find

$$\sum_{n=-\infty}^{\infty} [(1 + e^{-2i\alpha})u_{nx} - i(1 - e^{-2i\alpha})u_{ny}]e^{in\alpha} = 2ae^{-i\alpha}.$$

Collecting the coefficients of the same basis components $e^{in\alpha}$ in the left-hand side, we have

$$\sum_{n=-\infty}^{\infty} (u_{nx} - iu_{ny} + u_{n+2\,x} + iu_{n+2\,y})e^{in\alpha} = 2ae^{-i\alpha}.$$

Thus, we obtain the infinite system of equations for the coefficients u_n:

$$u_{-1\,x} - iu_{-1\,y} + u_{1\,x} + iu_{1\,y} = 2a, \qquad (A.5)$$

$$u_{nx} - iu_{ny} + u_{n+2\,x} + iu_{n+2\,y} = 0, \qquad n \neq -1. \qquad (A.6)$$

From formula (4) it follows that for the unique determination of the function u it suffices to find the complex vector $u = (u_0, u_1, u_2, \dots) \in l_2$. Setting $u = v + iw$, where v and w are real vectors from l_2, and separating the real and imaginary parts in (6) for $n = 0, 1, 2, \dots$, we obtain the following system:

$$(E - A)v_y = (E + A)w_x, \qquad (A.7)$$

$$(E + A)v_x = -(E - A)w_y. \qquad (A.8)$$

Here E is the unit operator and $A = (U^*)^2$, where U is the operator of the shift to the right:

$$U(u_0, u_1, u_2, \dots) = (0, u_0, u_1, \dots).$$

Appendix A

Applying the operator $(E - A)\partial_y$ to (7) and the operator $(E + A)\partial_x$ to (8), after summing up we get

$$\Delta_A v \equiv (E + A)^2 v_{xx} + (E - A)^2 v_{yy} = 0.$$

The equation for the imaginary part $\Delta_A w = 0$ is obtained analogously. From our construction and the fact that the vector function u is given on the boundary $\partial\Omega$:

$$u|_{\partial\Omega} = f = (f_0, f_1, \dots),$$

where f_n are the Fourier coefficients of f, it follows that the original problem (1), (2) is reduced to the solution of two Dirichlet problems for the operator Δ_A:

$$\Delta_A v = 0, \qquad v|_{\partial\Omega} = g,$$
$$\Delta_A w = 0, \qquad w|_{\partial\Omega} = h, \qquad (A.9)$$

Here g and h are the real and imaginary parts of the vector $f = g + ih$, respectively. Formulas (4), (5), and the representation $u = v + iw$ imply that the unknown function is as follows:

$$a(x, y) = v_{1x} - w_{1y}. \qquad (A.10)$$

We show that the function a can be found by v only, i.e., by the solution of the problem (9). Multiplying (7) by $(E - A)^{-1}$, we obtain

$$w_y = (E - A)^{-1}(E + A)v_x.$$

Therefore,

$$v_x - w_y = 2(E - A)^{-1} v_x. \qquad (A.11)$$

The operators $E \pm A$ have unbounded inverse operators. This follows from the equalities below, which can be easily verified:

$$U^*(f_0, f_1, \dots) = (f_1, f_2, \dots),$$

$$\operatorname{Sp} U^* = \{\lambda \in \mathbb{C} \mid |\lambda| < 1\}.$$

Here $\operatorname{Sp} U^*$ is the discrete spectrum of the operator U^*. Identifying the function a with the vector $\boldsymbol{a} = (a, 0, 0, \dots)$, from (10), (11), and the definition of A we obtain

$$\boldsymbol{a} = 2(E - UU^*)U^*(E - U^*)^{-1}(E + U^*)^{-1} v_x. \qquad (A.12)$$

The formula which expresses the function a in terms of w is obtained analogously. When deducing (12) we have used the identity

$$(E - UU^*)U^*f = (f, 0, 0, \dots), \qquad f \in l_2.$$

We now prove that the function v and the gradient of the function w are uniquely determined from the system (7), (8) and the boundary condition $v|_{\partial\Omega} = g$. To this end, first, we prove the uniqueness of determination of the right-hand side of the equation $Pu = a$ by the even part f_+ of the function f with respect to α. We set $\nu^\perp = (-\sin\alpha, \cos\alpha)$. In view of the identity

$$2\langle \nu^\perp, \nabla u\rangle \partial_\alpha \langle \nu, \nabla u\rangle = |\nabla u|^2 + \partial_\alpha \{\langle \nu^\perp, \Delta u\rangle \langle \nu, \Delta u\rangle\}$$
$$+ (u_y u_\alpha)_x - (u_x u_\alpha)_y, \qquad (A.13)$$

after integrating along $\Omega \times [-\pi, \pi]$, applying the Stokes formula and observing that $\partial_\alpha \langle \nu, \nabla u\rangle = \partial_\alpha a(x, y) = 0$, we have

$$\int_\Omega \int_{-\pi}^\pi |\nabla u|^2 \, dx \, dy \, d\alpha = -\int_{\partial\Omega} \int_{-\pi}^\pi u_\alpha u_\tau \, d\alpha. \qquad (A.14)$$

Here $u_\tau = \langle \tau, \nabla u\rangle$, τ is a unit tangent vector to Ω, ds is the length element of the curve $\partial\Omega$. Set $u_\pm = (u(x, y, \alpha) \pm u(x, y, -\alpha))/2$. Then we have

$$u_\alpha u_\tau = (u_+)_\alpha (u_+)_\tau + (u_-)_\alpha (u_-)_\tau + (u_+)_\alpha (u_-)_\tau + (u_-)_\alpha (u_+)_\tau.$$

The first two terms vanish after being substituted into the right-hand side of equality (14) since they are odd. For the remaining integrals, taking into account the ε-inequality $2xy \leq \varepsilon x^2 + \varepsilon^{-1} y^2$, $\varepsilon > 0$, we obtain

$$2\int_\Omega \int_{-\pi}^\pi |\nabla u|^2 \, dx \, dy \, d\alpha \leq \varepsilon \int_{\partial\Omega} \int_{-\pi}^\pi (|u_{-\tau}|^2 + |u_{-\alpha}|^2) \, ds \, d\alpha$$
$$+ \frac{1}{\varepsilon} \int_{\partial\Omega} \int_{-\pi}^\pi (|u_{+\tau}|^2 + |u_{+\alpha}|^2) \, ds \, d\alpha. (A.15)$$

Setting $u_+|_{\partial\Omega} = 0$ in this estimate and passing to the limit as $\varepsilon \to 0$, we obtain

$$\nabla u = 0 \quad \Rightarrow \quad a = 0.$$

Since

$$u_+ = \sum_{n=-\infty}^\infty v_n e^{in\alpha}, \qquad u_- = i \sum_{n=-\infty}^\infty w_n e^{in\alpha},$$

we obtain estimate (15) in terms of v and w:

$$\|\nabla v\|^2 \leq \varepsilon(\|w_\tau\|^2 + \|\mathcal{D}w\|^2) + \varepsilon^{-1}(\|g_\tau\|^2 + \|\mathcal{D}g\|^2). \tag{A.16}$$

Here

$$\mathcal{D}g = (0, g_1, 2g_2, \ldots, ng_n, \ldots), \qquad \|g\|^2 = \sum_{n=0}^{\infty} \|g_n\|_{L_2(\Omega)}^2.$$

Hence, for $g = 0$ we have $v = 0$ and $\nabla w = 0$ in view of (7), (8).

Projecting the operator Δ_A or the operator that generates the system (7), (8) onto the subspace $l_{2N} = \{u = (u_0, u_1, \ldots, u_N, 0, \ldots)\}$ and solving the obtained finite elliptic problem by means of formula (10) or (12), we find an approximate solution of the original problem. The corresponding convergence estimates follow from (15) and (16). Since there are analogues of identity (13) for equations of more general type, for example, for

$$\langle \nu, \nabla u \rangle + k(x, y, \alpha)u_\alpha + d(x, y, \alpha)u = a(x, y),$$

$$\langle \nu, \nabla u \rangle + \langle \nu_\alpha, \nabla \ln a \rangle u_\alpha + 2a(x, y) = 0, \qquad a(x, y) > 0.$$

the results obtained above can be generalized to the corresponding boundary value problems for these equations.

The approach described here was proposed by the author in Bukhgeim (1982). Identity (13) and its nonlinear analogues were first obtained by Mukhometov (1977), who used these identities to prove the uniqueness theorems and stability estimates for a wide range of problems of the form (1), (2). In the class of analytic functions, the inverse problem (1), (2) and its nonlinear analogues were first studied by Yu. E. Anikonov (see Anikonov (1978) and the bibliography therein). In this case, f was given on a part of the boundary $Sigma$. Further references can be found in Romanov (1984). The detailed presentation of the Radon problem and its generalizations together with the original Radon's work (of 1917) is given in Helgason (1980).

Bibliography

Alekseev, A. S. (1967). Inverse dynamic problems of seismology. In: *Some Methods and Algorithms of Interpretation of Geophysical Data*. Nauka, Moscow, 9–84 (in Russian).

Alekseev, A. S. and Dobrinskii, V. I. (1975). Some problems of practical use of inverse dynamic problems of seismology. In: *Mathematical Problems of Geophysics. Vol. 6, No. 2*. Computing Centre, Siberian Branch of the USSR Acad. Sci., Novosibirsk, 7–53 (in Russian).

Androshchuk, A. A. (1986). On the uniqueness of reconstruction of the two-dimensional Schrödinger equation. *Dokl. Akad. Nauk SSSR* 291 (5), 1033–1036 (in Russian).

Anikonov, Yu. E. (1978). *Some Methods of Study of Multidimensional Inverse Problems for Differential Equations*. Nauka, Novosibirsk (in Russian).

Apartsin, A. S. and Bakushinskii, A. B. (1972). Approximate solution of Volterra integral equations of the first kind by the method of quadrature sums. In: *Differential and Integral Equations. Vol. 1*. Irkutsk, 248–258 (in Russian).

Atkinson, F. (1964). *Discrete and Continuous Boundary Problems*. Academic Press, New York.

Bakushinskii, A. B. (1971). Difference schemes for the solution of ill-posed abstract Cauchy problems. *Diff. Uravn.* 7 (10), 1876–1885 (in Russian).

Bakushinskii, A. B. (1972). On the solution of the ill-posed Cauchy problem for an abstract differential equation of the second order by difference methods. *Diff. Uravn.* 8 (5), 881–890 (in Russian).

Berezanskii, Yu. M. (1958). On the uniqueness theorem for the inverse problem of spectral analysis for the Schrödinger equation. *Trudy Mosk. Math. Obshch.* 7, 3–51, Moscow (in Russian).

Berezanskii, Yu M. (1965). *Expansion of Self-Adjoint Operators with respect to Eigenfunctions.* Naukova Dumka, Kiev (in Russian).

Berezanskii, Yu. M. (1985). Integration of nonlinear difference equations by the method of inverse problem of spectral analysis. *Dokl. Akad. Nauk SSSR,* 281 (1), 16–19 (in Russian).

Beznoshchenko, N. Ya. and Prilepko, A. I. (1977). Inverse problems for equations of parabolic type. In: *Problems of Math. Phys. and Comput. Math.* Moscow, 51–63 (in Russian).

Bukhgeim, A. L. (1972). On a class of Volterra operator equations of the first kind. *Functional Analysis and its Applications* 6 (1), 1–9 (in Russian).

Bukhgeim, A. L. (1982). Inverse problems and integral equations of special form. In: *Methods of Solution of Ill-Posed Problems and their Applications.* Computing Centre, Siberian Branch of the USSR Acad. Sci., Novosibirsk, 184–186 (in Russian).

Bukhgeim, A. L. (1983a). On the stability of difference schemes for ill-posed problems. *Dokl. Akad. Nauk SSSR* 270 (1), 26–28 (in Russian).

Bukhgeim, A. L. (1983b). *Volterra Equations and Inverse Problems.* Nauka, Novosibirsk (in Russian).

Bukhgeim, A. L. (1984). Carleman estimates for Volterra operators and the uniqueness in inverse problems. *Sib. Mat. Zh.* 25 (1), 53–60 (in Russian).

Bukhgeim, A. L. (1985). Multidimensional inverse problems of spectral analysis. *Dokl. Akad. Nauk SSSR* 284 (1), 21–24 (in Russian).

Bukhgeim, A. L. (1986a). *Difference Methods of Solution of Ill-Posed Problems.* Computing Centre, Siberian Branch of the USSR Acad. Sci., Novosibirsk (in Russian).

Bukhgeim, A. L. (1986b). Integral equations in diagnosis problems. *Electronic Modelling* 198, 54–56 (in Russian).

Bukhgeim, A. L. and Kardakov, V. B. (1978). Solution of the inverse problem for the elastic wave equation by the method of spherical means. *Sib. Mat. Zh.* 19 (4), 749–757 (in Russian).

Bukhgeim, A. L. and Klibanov, M. V. (1981). Uniqueness in the whole for a class of multidimensional inverse problems. *Dokl. Akad. Nauk SSSR* 260 (2), 269–272 (in Russian).

Bukhgeim, A. L. and Yakhno, V. G. (1976). On two inverse problems for differential equations. *Dokl. Akad. Nauk SSSR* 229 (4), 785–786 (in Russian).

Calderon, A. P. (1959). Uniqueness in the Cauchy problem for partial differential equations. *Amer. J. Math.* 26, 16–36.

Carleman, T. (1939). Sur un probléme d'unicité pour les systémes d'équations aux dérivées partielles á deux variables indépendentes. *Arkiv Math. Astr. Fys.* 26 (17), 1–9.

Chadan, K. and Sabatier, P. (1977). *Inverse Problems in Quantum Scattering Theory*. Springer-Verlag, New York.

Chudov, L. A. (1962). Difference methods of solution of the Cauchy problem for the Laplace equation. *Dokl. Akad. Nauk SSSR* 143 (4), 798–801 (in Russian).

Chudov, L. A. (1967). Difference schemes and ill-posed problems for partial differential equations. *Vych. met. i prog.* 8, 34–62 (in Russian).

Courant, R. (1962). *Partial Differential Equations. Vol. II. Methods of Mathematical Physics*. Interscience Publishers, New York, London.

Dunford, N. and Schwartz, J. T. (1971). *Linear Operators. Vol. III. Spectral Operators*. Interscience Publishers, New York.

Delsarte, J. (1938). Sur certaines transformation fonctionelles relative aux equations lineaires aux derivees. *C.R. Acad. Sci. Ser.* 206, 178–182.

Dobrinskii, V. I. and Lavrent'ev, M. M. (1978). The way of determination of the profile of sound velocity by recording the acoustic field in a single point. In: *Propagation of Acoustic Waves*. Vladivostok, 12–14 (in Russian).

Douglas, J. and Gallie T. M. (1959). An approximate solution of an improper boundary value problem. *Duke Math. J.* 26, 339–348.

Faddeev, L. D. (1974). The inverse problem of the quantum scattering theory. In: *Modern Problems of Mathematics. Vol. 3*. Moscow, VINITI, 93–180 (in Russian).

Fichera, G. (1974). *Existence Theorems in Elasticity Theory.* Mir, Moscow (in Russian)

Freeman, J. M. (1965). Volterra operator similar to $I : f \to \int_0^x f(t)\,dt$. *Trans. Amer. Math. Soc.* 116, 181–192.

Garipov, R. M. and Kardakov, V. B. (1973). The Cauchy problem for the wave equation with nonspatial initial manifold. *Dokl. Akad. Nauk SSSR* 213 (5), 1047–1050 (in Russian).

Gel'fand, I. M. and Shilov, G. E. (1958). *Generalized Functions and Operations on them.* Fizmatgiz, Moscow (in Russian).

Gel'fond, A. O. (1967). *Finite-Differences Calculus.* Nauka, Moscow (in Russian).

Hadamard, J. (1932). *Le Problème de Cauchy et les Équations aux Dérivées Partiélles Linéaires Hyperboliques.* Hermann, Paris.

Helgason, S. (1980). *The Radon Transform.* Birkhäuser, Boston.

Hörmander, L. (1969). *Linear Partial Differential Operators.* Springer-Verlag, Berlin, Heidelberg, New York.

Hörmander, L. (1990). *The Analysis of Linear Partial Differential Operators. Differential Operators with Constant Coefficients.* Springer-Verlag, Berlin, New York.

Isakov, V. M. (1980). On the uniqueness of the solution of the Cauchy problem. *Dokl. Akad. Nauk SSSR* 255 (1), 18–21 (in Russian).

Ivanov, V. K. (1962). On linear ill-posed problems. *Dokl. Akad. Nauk SSSR* 145 (2), 270–272 (in Russian).

Ivanov, V. K., Vasin, V. V., and Tanana, V. P. (1978). *Theory of Linear Ill-Posed Problems and its Applications.* Nauka, Moscow (in Russian).

John, F. (1955). Numerical solution of the heat equation for preceding time. *Ann. Mat. Pure Appl.* 40, 129–142.

Klibanov, M. V. (1984a). Uniqueness in the whole in inverse problems for a certain class of differential equations. *Diff. Uravn.* 20 (1), 1947–1954 (in Russian).

Klibanov, M. V. (1984b). Inverse problems in the whole and the Carleman estimates. *Diff. Uravn.* 20 (6), 1035–1042 (in Russian).

Klibanov, M. V. (1985). A class of inverse problems for linear parabolic equations. *Dokl. Akad. Nauk SSSR* 280 (3), 533–536 (in Russian).

Krasnosel'skii, M. A. and Zabreiko, P. P. (1975). *Geometric Methods of Nonlinear Analysis*. Nauka, Moscow (in Russian).

Krein, S. G. (1971). *Linear Equations in a Banach Space*. Nauka, Moscow (in Russian).

Krein, S. G. and Prozorskaya, O. I. (1963). On approximate methods of solution of ill-posed problems. *Zh. Vych. Mat. i Mat. Fiz.* 3 (1), 120–130 (in Russian).

Kunetz, G. (1963). Quelques examples l'analyse d'enregistrements sismiques. *Geophys. Prospect* 11 (4), 87–96.

Ladyzhenskaya, O. A. (1953). *Mixed Problem for the Hyperbolic Equation*. Gostekhizdat, Moscow (in Russian).

Lattès, R. and Lions, J.-L. (1967). *Méthode de Quasi-réversibilité et Applications*. Dunod, Paris.

Lavrent'ev, M. M. (1962). *On Some Ill-Posed Problems of Mathematical Physics*. Nauka, Novosibirsk (in Russian).

Lavrent'ev, M. M. (1965). On a class of inverse problems for differential equations. *Dokl. Akad. Nauk SSSR* 160 (1), 32–35 (in Russian).

Lavrent'ev, M. M., Reznitskaya, K. G., and Yakhno, V. G. (1982). *One-Dimensional Inverse Problems of Mathematical Physics*. Nauka, Novosibirsk (in Russian).

Lavrent'ev, M. M., Romanov, V. G., and Shishatskii, S. P. (1980). *Ill-Posed Problems of Mathematical Physics and Analysis*. Nauka, Moscow (in Russian).

Leibenzon, Z. L. (1966). Inverse problem of spectral analysis of ordinary differential operators of higher orders. *Trudy Mosk. Mat. Obshch.* 15, 70–144 (in Russian).

Levitan, B. M. (1962). *Operators of Generalized Shift*. Nauka, Moscow (in Russian).

Levitan, B. M. (1984). *Inverse Sturm-Liouville Problems.* Nauka, Moscow (in Russian).

Marchenko, V. A. (1950). Some problems of the theory of differential operators of the second order. *Dokl. Akad. Nauk SSSR* 72 (3), 457–460 (in Russian).

Marchenko, V. A. (1972). *Spectral Theory of the Sturm-Liouville Operator.* Naukova Dumka, Kiev (in Russian).

Marchenko, V. A. (1977). *The Sturm-Liouville Operators and their Applications.* Naukova Dumka, Kiev (in Russian).

Marchuk, G. I. (1980). *Methods of Computational Mathematics.* Nauka, Moscow (in Russian).

Maurin, K. (1959). *Metody przestrzeni Hilberta.* Panstwowe Wyd. Naukowe, Warsaw.

Morozov, V. A. (1966). On regularization of ill-posed problems and the choice of the regularization parameter. *Zh. Vych. Mat. i Mat. Fiz.* 6 (1), 170–175 (in Russian).

Morozov, V. A. (1974). *Regular Methods of Solution of Ill-Posed Problems.* Moscow State University, Moscow (in Russian).

Mukhometov, R. G. (1977). The problem of reconstruction of the two-dimensional Riemann metric and integral geometry. *Dokl. Akad. Nauk SSSR* 232 (11), 32–35 (in Russian).

Nirenberg, L. (1975). Lectures on linear partial differential equations. *Uspekhi Mat. Nauk,* 30 (4), 147–204 (in Russian).

Nirenberg, L. (1977). *Lectures on Nonlinear Functional Analysis.* Mir, Moscow (in Russian).

Nizhnik, L. P. (1973). *Inverse Nonstationary Scattering Problem.* Naukova Dumka, Kiev (in Russian).

Ovsyannikov, L. V. (1965). Singular operator in a scale of Banach spaces. *Dokl. Akad. Nauk SSSR* 163 (4), 819–822 (in Russian).

Ovsyannikov, L. V. (1971). Nonlinear Cauchy problem in a scale of Banach spaces. *Dokl. Akad. Nauk SSSR* 200 (4), 789–792 (in Russian).

Payne, L. E. (1973). Some general remarks on improperly posed problems for partial differential equations. In: *Lecture Notes in Mathematics. Vol. 316.* Springer–Verlag, Berlin, 1–30.

Phillips, D. L. (1962). A technique for the numerical solution of certain integral equations of the first kind. *J. Assoc. Comput. Nach.* 9 (1), 84–97.

Pólya, G. and Szegö, G. (1964). *Aufgaben und Lehrsätze aus der Analysis. Vol. II.* Springer-Verlag, Berlin, New York.

Povzner, A. Ya. (1948). On differential equations of the Sturm-Liouville type on a semiaxis. *Mat. Sb.* 23 (1), 3–52 (in Russian).

Preobrazhenskii, N. G. and Pikalov, V. V. (1982). *Unstable Problems of Plasma Diagnosis.* Nauka, Novosibirsk (in Russian).

Prony, G. R. B. (1975). Essai experimental et analytique, etc. *I. de L'Ecole Polytechnique* 1 (2), 24–76.

Pucci, C. (1955). Sui problemi di Cauchy non "ben posti". *Rend. Asad. Naz. Lincei* 18, 473–477.

Riesz M. (1938). Integrales de Riemann–Lionville et potentiels. *Acta Szeged* 9, 1–42.

Romanov, V. G. (1973). One-dimensional inverse problem for the wave equation. *Dokl. Akad. Nauk SSSR* 211 (5), 1083–1084 (in Russian).

Romanov, V. G. (1984). *Inverse Problems of Mathematical Physics.* Nauka, Moscow (in Russian).

Samarskii, A. A. (1977). *Theory of Difference Schemes.* Nauka, Moscow (in Russian).

Selected Problems (1977). (Ed. V. M. Alekseev). Mir, Moscow (in Russian).

Stieltjes, T. J. (1936). *Research on Continued Fractions.* ONTI, Moscow (in Russian).

Taylor, M. (1981). *Pseudodifferential Operators.* Princeton University Press, Princeton.

Ten Men Yan (1984). Linear multistep methods for numerical solution of the Volterra equation of the first kind. In: *Optimization Methods and Operations Research.* Irkutsk, 254–257 (in Russian).

Tikhonov, A. N. (1943). On the stability of inverse problems. *Dokl. Akad. Nauk SSSR*, 39 (5), 195–198 (in Russian).

Tikhonov, A. N. (1963). On solution of ill-posed problems and the regularization method. *Dokl. Akad. Nauk SSSR* 151 (3), 501–504 (in Russian).

Tikhonov, A. N. and Arsenin, V. Ya. (1979). *Methods of Solution of Ill-Posed Problems*. Nauka, Moscow (in Russian).

Treves J. F. (1967). On the theory of linear partial differential operators with analytic coefficients. *Trans. Amer. Math. Soc.* 137, 1–20.

Vabishchevich, P. N., Golovizin, V. M., Elenin, G. G., *et al.* (1986). *Computation Methods in Mathematical Physics*. Moscow State University, Moscow (in Russian).

Vladimirov, V. S. (1971). *Equations of Mathematical Physics*. Nauka, Moscow (in Russian).

Voronin, V. V. and Tsetsokho, V. A. (1981). Numerical solution of integral equations with logarithmic singularity by the method of interpolation and collocation. *Zh. Vych. Mat. i Mat. Fiz.* 21 (1), 40–53 (in Russian).

Zakharov, B. N. and Suz'ko, A. A. (1981). *Potentials and the Quantum Scattering: Direct and Inverse Problem*. Energoatomizdat, Moscow (in Russian).